ELECTROMAGNETISM AND QUANTUM THEORY

ELECTRICAL SCIENCE
A Series of Monographs and Texts

Editors: Henry G. Booker and Nicholas DeClaris
A complete list of titles in this series appears at the end of this volume

ELECTROMAGNETISM AND QUANTUM THEORY

Dale M. Grimes

DEPARTMENT OF ELECTRICAL ENGINEERING
UNIVERSITY OF MICHIGAN
ANN ARBOR, MICHIGAN

1969

ACADEMIC PRESS New York and London

ACADEMIC PRESS, INC.
111 Fifth Avenue, New York, New York 10003

United Kingdom Edition published by
ACADEMIC PRESS, INC. (LONDON) LTD.
Berkeley Square House, London W1X 6BA

LIBRARY OF CONGRESS CATALOG CARD NUMBER: 79-91419

PRINTED IN THE UNITED STATES OF AMERICA

PREFACE

This book presents a view of quantum theory on the basis of a specific model: electromagnetic field theory. The idea for this presentation originated from two separate courses taught to first- or second-year graduate students in the Department of Electrical Engineering at The University of Michigan: "Electromagnetic Field Theory III" and "Quantum and Statistical Electronics." A way of treating both topics as one resulted. This point of view emphasizes the electromagnetic foundations of atomic phenomena, rather than their mechanical analogies.

Current quantum theory is the result of many years of evolution. One who looks back at the evolutionary process can see that each step has resulted from the exigencies of a particular situation, but he cannot help judging each conclusion as it does or does not seem to him to have nudged the process in the right direction. Furthermore, he himself, although he may be trying to synthesize a theory, is at one more stage in its evolution, and cannot be sure whether that stage will serve as a stable platform for further steps in the evolutionary process, or whether it is only a useful illusion, soon to fall away. All he can do is examine the evidence, old and new, one more time.

In writing this book, we begin from first principles and construct a physical model in an attempt to reconcile the electromagnetic theory with those of both classical and quantum mechanics. Some of the ideas presented are commonly accepted. Some have been suggested but not widely accepted, and some are new. We do not stop to categorize each one as we present it; rather, we strive for clarity, consistency, and continuity of argument, developed with reasonable rigor.

We do not expect everyone to agree with all that is presented here, for there is one main point we do not—we cannot—prove, that regarding the unradiated energy necessarily accompanying an oscillator. On another point, however, all can agree: many knowledgable persons disagree about the basis of quantum theory. We believe that the basic idea of this book is sound. The unifying aspects of electromagnetism and quantum theory have gone unstressed for too long. Moreover, most thinking on the relationship between the theories has proceeded without reference to physical models. But there are so many possibilities, and the physical significance of the mathematics is so remote, that a physical model of some sort, we believe, will sooner or later prove essential.

Even though an attempt to rebuild a physical theory upon currently unaccepted postulatory bases inspires in many men about the same reaction as the taste of vinegar in their wine, we believe that atomic physics is not a closed subject and that continuing attempts to restructure it are necessary.

DALE M. GRIMES

Ann Arbor
August 1969

ACKNOWLEDGMENTS

This book represents continuing hours of thought throughout more years than I like to admit. During that period of time, I have been generously supported and abetted by many. At least one, and in most cases several, generations of graduate students have studied and criticized each chapter. My departmental colleagues have been generous with their time and thought. Three different departmental chairmen have aided me with material support. But most of all I am indebted to my wife. The book could not have reached fruition had she not been willing, time and again, to tolerate an inordinately preoccupied husband.

CONTENTS

Preface v

Acknowledgments vii

A Table of Physical Constants xv

INTRODUCTION 1

Chapter 1: THE LORENTZ TRANSFORMATIONS 7

In this chapter, the concepts of the cosmological principle and the relativity principle are developed. Using them, the Lorentz transformations are in turn developed, as are some of their more elementary consequences.

Chapter 2: ELECTROMAGNETIC TENSORS 15

The four-dimensional formulation of tensors that will later be shown to incorporate the laws of electrodynamics are formulated and interrelated. The three-dimensional forms are recognized. Maxwell's equations result.

Chapter 3: EVALUATION OF THE FOUR-POTENTIAL 23

The electromagnetic tensors defined in Chapter 2 are known only in terms of a differential equation. Full knowledge of the system requires the integration of that equation. For that purpose, the four-dimensional Green's function is developed here.

Chapter 4: ELECTRODYNAMICS 29
The work of the three previous chapters is pulled together in this chapter. The significance of the field vectors is developed, as are Poynting's theorem and Newton's laws.

Chapter 5: MULTIPOLAR FIELDS 41
A general expansion for electromagnetic fields, using spherical coordinates, is developed in this chapter.

Chapter 6: ENERGY STORAGE AND FLOW IN MULTIPOLAR FIELDS 51
The power, momentum, and angular momentum flow through a surface are found using the expansions of Chapter 5. Then the power flow expression is used to find the complex radiation reaction upon a source. The radiant Q is defined and radiant properties analysed as a function of Q. Many conclusions are then drawn regarding radiation from accelerating electrons.

Chapter 7: ENERGY AND MOMENTUM IN MULTIPOLAR FIELDS 63
In this chapter, the net radiated power, momentum, and angular momentum are found in terms of field parameters. An incoming plane wave is next added and the boundary conditions met. The result is a nearly complete expression for the radiated dynamic properties. One boundary condition is that the ratio of radiated energy to momentum be c. A result of that condition is that radiated energy and angular momentum are in the ratio of the

angular frequency. The first condition is commonly taken to require quantized radiation. The second condition is well known but thought to arise only from quantum mechanical processes.

Chapter 8: THE ELECTRON 75

The known properties of an electron are discussed and tabulated. Next we show that two opposite, paired, and isolated charges, with the appropriate energy, will spiral into each other and should therefore be unstable. However, we next show that charge configurations do exist which are stable in the sense that no power is radiated into the far field. We solve for several such possible charge distributions and then show that others exist as well.

Chapter 9: AN ERGODIC EQUATION 87

We take the results of Chapter 8 to show that stable charge configurations can occur in isolation which are in equilibrium with their own reactive field, as discussed in Chapters 5–7. We then assume that some type of stable charge configuration exists and treat the problem as a thermodynamic one. The result is that the charge configuration obeys Schrödinger's equation, with Planck's constant as an undetermined parameter. Then we treat the same problem in accordance with the techniques of statistical mechanics, and a model of a point charge being driven by a reactive field. Once again the result is Schrödinger's equation, but this time Planck's constant is given in terms of other parameters.

Chapter 10: PROPERTIES OF THE WAVE FUNCTION 97

Given the results of Chapter 9, the general characteristics of charged regions are found.

These include methods of solving for both dynamic and configurational properties of the charged region. The Heisenberg uncertainty principle and the Pauli exclusion principle result.

Chapter 11: A TIME-DEPENDENT BUT ERGODIC EQUATION 107
The time-dependent Schrödinger equation is developed.

Chapter 12: THE RADIATING ATOM 111
The Manley–Rowe equations are developed and applied to atomic processes. A result is that the ratio of energy flows is equal to the ratio of the frequencies. This result combines with those of Chapter 11 to show that the ratio of energy to frequency is Planck's constant, and with the results of Chapter 7 to show that angular momentum is exchanged in units equal to Planck's constant.

Chapter 13: THE HARMONIC POTENTIAL 117
Schrödinger's equation is solved for the case of a harmonic potential.

Chapter 14: A SPHERICAL POTENTIAL 121
Schrödinger's equation is solved for the case of a spherically symmetric potential.

Chapter 15: THE COULOMB POTENTIAL 125
Schrödinger's equation is solved for the case of a Coulomb potential. The result is the quantum-mechanical solution of a hydrogen atom.

Chapter 16: TRANSITION PROBABILITIES 131

The results of Chapters 11 and 15 are combined to solve for the parameters of dipolar atomic radiation. These parameters agree with those of Chapters 7 and 12, and contain expressions for transitional probabilities as well.

Chapter 17: ANGULAR TORQUE ON DYNAMIC SYSTEMS 139

The electric and magnetic fields generated by a closed current loop are found. These solutions are then extended to the case of *n* separate but equally spaced charges. Then two concentric loops with equal numbers of charges and rotating with equal peripheral speeds are considered. The result is a stable configuration that satisfies the rules of space quantization, hitherto thought to be a purely quantum-mechanical phenomenon.

Subject Index 147

A Table of Physical Constants

Speed of light	c	$= 2.998 \times 10^{8}\ \mathrm{ms}^{-1}$
Permeability of vacuum	μ	$= 4\pi \times 10^{-7}\ \mathrm{Hm}^{-1}$
		$= 12.637 \times 10^{-7}\ \mathrm{Hm}^{-1}$
Permittivity of vacuum	$\varepsilon = 1/c^2\mu$	$= 8.8543 \times 10^{-12}\ \mathrm{Fm}^{-1}$
Impedance of vacuum	$\eta = \mu c$	$= 378.84\ \mathrm{Js}\ \text{Å}^{-2}$
Planck's constant	$\hbar$	$= 1.0544 \times 10^{-34}\ \mathrm{Js}$
Fine-structure constant	$\alpha = \eta e^2/4\pi\hbar$	$= 7.297 \times 10^{-3}$
First Bohr orbit	$b = \hbar/\alpha mc$	$= 5.292 \times 10^{-11}\ \mathrm{m}$
Electron charge	e	$= 1.602 \times 10^{-19}\ \mathrm{C}$
Electron rest mass	m	$= 9.108 \times 10^{-31}\ \mathrm{kg}$
Electron magnetic moment		$= 0.9284 \times 10^{-23}\ \text{Å}\ m^2$
		$= 1.0011\ e\hbar/2m$
Lorentz diameter, electron D_1	$= \mu e^2/4\pi m$	$= 2.818 \times 10^{-15}\ \mathrm{m}$
Compton wavelength, electron	$2\pi\hbar/mc$	$= 2.426 \times 10^{-12}\ \mathrm{m}$
Proton rest mass	M	$= 1.672 \times 10^{-27}\ \mathrm{kg}$
Proton magnetic moment		$= 1.4104 \times 10^{-26}\ \text{Å}\ \mathrm{m}^2$
		$= 2.793\ e\hbar/2M$
Lorentz diameter, proton		$= 1.5347 \times 10^{-18}\ \mathrm{m}$
Compton wavelength, proton		$= 1.321 \times 10^{-15}\ \mathrm{m}$
Measured diameter, proton		$= 1.81 \times 10^{-15}\ \mathrm{m}$

ELECTROMAGNETISM AND QUANTUM THEORY

INTRODUCTION

Our analysis begins with the observation that the science of electricity and magnetism is, in several ways, all that a science should be. It comprises a vast complex of information that can be understood both qualitatively and quantitatively, and that has made it commonplace to predict events by solving the applicable partial differential equations and then matching the boundary conditions. Furthermore, the equations basic to that science are firmly grounded on a few general postulates: that a scalar invariant charge exists, that the relationship between events in two inertial frames can be correctly described by a linear transformation, and that the speed at which electromagnetic energy propagates is constant when measured at two or more inertial locations at different times. Even better, the logical development of the laws is general enough to permit the *a priori* determination that they should apply from the largest linear dimension at least down to the diameter of the effective "point" charge.

On the atomic scale of dimensions, however, this ideal structure appeared to break down. Two point charges, such as an electron and an "infinite-mass" proton, cannot be held in equilibrium by electrostatic forces alone; for stability, then, there must be relative motion between them. In that case, at least one of them, say the electron, must accelerate, and therefore radiate energy; and if it radiates energy, it will spiral inexorably toward the nucleus. But atoms are stable; hence, it has long been concluded that the electrodynamic laws do not apply in atomic conditions. Thus a critical result of the laws of electromagnetism has appeared to be invalid well within the expected range of their validity.

The inadequacy of classical electrodynamics within the realm of the atom was pointed out in detail during the second decade of this century. In 1913 Bohr advanced arguments pertaining to energy exchange by atoms, and in 1917 Einstein advanced arguments based upon the absorption and emission of radiant energy and momentum by atoms. These arguments, in effect, have prevented attempts to apply classical theory within the realm of the atom. They also were the axiomatic base for the debate of the thirties, between the same two men, from which has come the now-orthodox view of quantum theory—a view never supported by Einstein.

According to Bohr, it was acknowledged by most scientists at the time of his early writings that classical electrodynamics was inadequate. As evidence, he stated that

(1) by classical theory, an electron in orbit about a nucleus can have no stable state;

(2) atomic structure can be explained only by artificially introducing a "quantity foreign to classical electrodynamics, i.e., Planck's constant"; and

(3) the frequency at which energy is absorbed and emitted by an atom in a given energy state is equal neither to the frequency associated with that state nor to any multiple of it.

Einstein found that the Planck radiation law imposes severe momentum requirements on the exchange of radiant energy by atoms. These conditions are

(1) the momentum exchanged in emission is equal in magnitude and opposite in direction to that exchanged in absorption; and

(2) the ratio of the energy to the momentum exchanged is c.

He concluded that, according to the classical laws, this ratio is correct for absorption from a plane wave but not for emission, and from that conclusion he inferred that such laws are inadequate.

In the studies of the structure of the atom which dominated physics in the first third of this century, these conclusions aroused much controversy. Many tried to circumvent or refute them, but all of them failed. It was Bohr who proceeded on the assumption that accelerated electrons in an equilibrated atomic environment do not radiate, and thus made his first major contribution. Under critical prodding, chiefly from Einstein, he went on to become chiefly responsible for the interpretation that has become orthodox quantum theory. His interpretations were able to take into account not only the already observed phenomena but also the flood of new experimental results pertaining to atomic structure; hence they have come to be the orthodox view.

But accepting this view requires much more than denying the validity of the electromagnetic equations. As Einstein pointed out, it necessitates applying statistics to the performance of individual particles in such a way as to require rejection of the deterministic concept that every event is the unique result of preceding ones. This rejection is itself unique, but most of the physicists active in developing the theory have come to accept it as a necessary consequence of the equations. Many others, however, have taken exception. Indeed, Bohr is quoted as saying that only after two years of argument and propaganda was he able to convince Heisenberg and Bloch that the physical observations made the rejection mandatory. And this rejection was surely one reason why Einstein, shortly before his death, wrote to Professor Born, " . . . your quantum theory is not the real McCoy."

Turning to the presently orthodox view of quantum theory, in it the Schrödinger equation is given a role on a par with those of relativity, and supersedes the Maxwell equations as necessary. Now this approach is in many ways very reasonable. It is logically self-consistent, it is convenient for one trained in classical mechanics, and it follows the pattern of thought that led to the development of quantum theory. And the Schrödinger equation serves as a firm base, for, when it is modified to meet relativistic requirements, it leads, like the Maxwell equations, to quantitative conclusions that fit a wide range of experimental data. No reasonable doubt of its correctness remains, at least through first order. But it presents some practical difficulties, as well as denying the principle of determinacy. One is that, since it has no logical basis, there is no way of *a priori* determining the limits of its validity. Furthermore, since many years of work based upon it have led to neither an adequate particle theory nor a unified field theory, its logical structure may well be inadequate to support these extrapolations, although it is difficult to recognize lack of progress toward a poorly defined objective.

With respect to orthodox interpretation of quantum theory, how important is its interpretation as long as the mathematics is correct? Einstein is reputed to have said something to the effect that one should examine physicists' mathematics, but ignore the models they build to interpret them. However, the importance of a good physical model is continually demonstrated. A very famous example is Maxwell's eminently successful formulation of electrodynamics, which differs from earlier mathematical formulations mainly in being based upon Faraday's model of lines of force. More mundanely, meaningful approximations are often made possible only by the use of physical reasoning to guide the selection of idealized models for analysis, to provide a basis for choosing a particular approximation method, and to give meaning to the coefficients in a power-series expansion. But who can say he knows the "true" interpretation of anything? In our culture, an interpretation is judged by its consistency, its simplicity, and its agreement with observation. The Babylonians provide an extreme example of the danger of the strict application of such rules. They developed a system to predict quantitatively the movements of the stars. All but seven—the sun, the moon, Mercury, Venus, Mars, Jupiter, and Saturn—were fixed in the firmament, which revolved about the earth-mountain once a day, under the guidance of the gods; the seven wanderers revolved, too, but had additional, predictable motions of their own. This theory meets all modern requirements. The difference is that the part of it which is not susceptible to experimental test does not satisfy modern standards; in talking of mechanical processes, we prefer guiding waves to guiding gods.

Is our present view of quantum theory on a firmer foundation than was the Babylonians' view of the universe? Is modern quantum theory a magnificent theoretical edifice that supports and perpetuates itself, but lacks a proper foundation? Although scientists of future generations will at least know if their edifice is similar to our own, each generation seems doomed to the temporally provincial belief that its own view is the correct one, and that earlier generations were naive.

The concepts upon which atomic physics is based have evolved but little since their inception, in spite of later, possibly relevant findings in other branches of science. Particles are still commonly assumed to be rigid, or deformation has been all but ignored, even though nuclear theory strongly suggests that under certain conditions particles are deformable. Moreover, from the vantage point of over half a century since Bohr and Einstein found it necessary to discard classical theory as inadequate to describe atomic phenomena, it appears that what they considered evidence for the breakdown of classical theory was rather due to the lack of certain mathematical techniques and experimental information that we now have.

REFERENCES

Bohm, David, "Causality and Chance in Modern Physics." Harper, New York, 1961.

Bohr, Niels, "On the constitution of atoms and molecules," *Phil. Mag.* **26,** 1–25 (1913).

Bohr, Niels, "Discussion with Einstein on epistemological problems in atomic physics," *in* "Albert Einstein, Philosopher-Scientist" (P. A. Schilpp, ed.), pp. 201–241. Harper, New York, 1959.

Born, Max, "Atomic Physics," 7th ed., Chap. 5. Hefner, New York, 1962.

Bronwell, A. B., "Is philosophy off limits?" *IEEE Spectrum,* **1,** 110–114 (1964).

de Broglie, L., "L'interpretation de la mecanique ondulatoire," *J. Phys. Radium* **20,** 963–979 (1959).

Dirac, P. A. M., "Foundations of quantum mechanics," *Nature* **203,** 115–116 (1964).

Einstein, Albert, "Remarks concerning the essays," *in* "Albert Einstein, Philosopher-Scientist" (P. A. Schilpp, ed.), pp. 665–688. Harper, New York, 1959.

Einstein, Albert, "The Quantum Theory of Radiation," *in* "The World of the Atom" (H. A. Boorse, and L. Motz, eds.), pp. 888–901. Basic Books, New York, 1966.

Feynman, R. P., "Space-time approach to nonrelativistic quantum mechanics," *Rev. Mod. Phys.* **20,** 367–387 (1948).

Koestler, Arthur, "The Sleepwalkers," pp. 20–21. Grosset & Dunlap (Universal Library Series), New York, 1963.

New York Times (*European ed.*), p. 1, July 2, 1965.

Nielson, J. R., "Memories of Niels Bohr," *Phys. Today* **16,** 22 (1963).

1

THE LORENTZ TRANSFORMATIONS

Since our environment has a physical extent, the idea of space is immediate and obvious to us; it is ingrained in our thought processes. And a physical body gives credence to the concept of distance by its own physical extent. A portion of the body can be considered separately and used as a unit by which the physical extents of other portions of the body can be measured. Thus the connection between the physical world and a geometry is easily made. Now two comparative measurements can be made only at different times. Furthermore, our environment changes continuously, and the changes occur in a definite order. A time scale can be established on the basis of the rate at which some event recurs; but any timing mechanism must have some physical extent. Therefore, the concepts of space and time cannot be completely separated.

The accepted views of the spacetime interrelationships are based on the theory of relativity: no way of extending them to very great or very small distances and times is yet known. Fortunately, for many purposes an understanding of space and time is not essential. The lack of it can be evaded by using properly chosen postulates, if those postulates are taken as describing the basis for physical events. When experimentally observed events show quantitative agreement with predictions based upon these postulates, the events are said to be understood. Newton's laws are examples of quantitative formulations of such postulates.

One postulate implicit in all of the sciences is the cosmological principle, that nature is not capricious: the universe exhibits no local irregularities, so that identical sets of circumstances produce identical results. One support of the cosmological principle, over a small range of distance and time, is that experiments conducted on earth are reproducible even though they must be conducted at different places and times, for the earth rotates on its axis and about the sun, the solar system moves through the galaxy, and time passes. Whether the cosmological principle applies over very great distance and time is not directly testable.

Let us apply this postulate to two separate but chemically similar bodies, say bodies *A* and *B*, which are the only two bodies in their universe and are so far apart that any interaction between them is negligible. A conclusion from the postulate is that the same physical laws, i.e., the same partial differential equations of motion, govern events on both. But it does not follow that the coefficients of each comparable term in those equations, i.e. the fundamental constants, are identical, for, by the same postulate, differences in local extensive variables can cause differences between constants on *A* and *B*. New units of length and time can be consistently defined on the basis of the physical extent of the atom, and the rate of recurring atomic events, on either body. Such units are natural units by which to determine time and distance and to measure both bodies. Now let us take the measured sizes of the two bodies, as separately determined on each by its own set of units, to be equal. Then, by the cosmological principle, the coefficient of each comparable term in the partial differential equations developed to describe the two bodies must be identical.

Now suppose that a pulse of radiation is emitted from *A* and moves past *B*. The speed of the pulse either is or is not a function of the circumstances that held on *A* when the pulse was emitted; terrestrial analogies would be an emitted projectile, whose velocity adds to that of its source velocity, and sound, whose velocity is independent of its source velocity. We are concerned here only with pulses whose speed does not depend upon the properties of the source. As such a pulse reaches that point in its course which is nearest *B*, a second pulse is emitted from the point on *B* nearest to that course, and moves in parallel with the pulse from *A*. If the speeds of both pulses are measured as they pass a second point on *B*, the result of the measurement will depend only upon the system of units on *B*; the measured speed *c* of the two pulses will be the same.

Now we repeat the experiment with relative motion between *A* and *B*. Since there are no other bodies in the universe, *A* can be considered moving and *B* fixed; hence, when the speed of the pulses is measured from *B*, *c* is functionally dependent only on local conditions at *B*, which are the same as when *A* was fixed. If our vantage point is reassigned to *A*, then *B* must be in motion; but, since no physical result can depend upon the method of calculation, the speed *c* is still independent of the speed of *B*. The effect of the requirement that *A* and *B* be identical depends upon how rapidly the fundamental constants vary with extensive parameters of the bodies.

Since other bodies can be brought into the universe without in any way affecting the foregoing argument, the spatial array of bodies in the universe can be constructed body by body without effect upon the conclusion that the speed of a pulse of radiation is independent of the relative speed, with respect to other bodies in the universe, of the body from which it is measured.

Let us see how to bring the known low-speed transformation laws into accord with this conclusion. We establish two coordinate systems C and C' with relative velocity v along the x axes. Suppose that an event is seen, and its position measured, from both systems. The result for system C, we call x; that for system C', x'. The problem is to relate the two coordinates x and x'. The simplest transformation between coordinate systems is linear and retains the low-velocity form $x' = (x - vt)$, as follows:

$$\begin{aligned} x' &= D(x - vt), \\ y' &= Ey, \\ z' &= Fz, \\ t' &= Gt - Hx, \end{aligned}$$

where D, E, F, G, and H are constants to be determined; in the limit of arbitrarily small velocities, each of the first four is equal to unity, and H is equal to zero.

A pulse of radiation is emitted from the origins of the two coordinate systems at the time when they coincide. Since the speed of radiation is c in both systems, the expanding wave front will be spherical in each system; that is, positions of constant phase satisfy the equalities

$$0 = -c^2t'^2 + x'^2 + y'^2 + z'^2,$$
$$0 = -c^2t^2 + x^2 + y^2 + z^2.$$

By using this condition and the linear transformation equations, we can eliminate the coordinates and obtain the constants. The result is

$$\begin{aligned} t' &= \gamma\left(t - \frac{vx}{c^2}\right), \qquad & y' &= y, \\ x' &= \gamma(x - vt), \qquad & z' &= z, \end{aligned} \tag{1-1}$$

where $\gamma^{-2} = 1 - (v^2/c^2)$. Equations (1-1) are the Lorentz transformation equations.

A rigid rod of length L_0 in the moving system will appear to a stationary observer to be shortened in the direction of its motion. Let L_0 be oriented in the direction of motion and fixed in the moving system. Photographs of the ends of the rod are taken from the stationary system at the same time t, and used to determine the length L of the rod. If $L_0 = x_2' - x_1'$, then, from Eq. (1-1),

$$L_0 = \gamma(x_2 - x_1) = \gamma L$$

or

$$L = \gamma^{-1} L_0, \tag{1-2}$$

and, since $\gamma \geqslant 1$, it follows that $L \leqslant L_0$, i.e., that the moving rod appears to be shorter than the stationary one. This result follows simply from a consideration of time and distance and so is independent of physical mechanisms inside the rod. The decrease in length is known as the Fitzgerald contraction.

A time interval also appears to be different according to whether it is measured in the moving frame or the stationary one. Suppose that a fixed time interval $T_0 = t_2' - t_1'$, such as a period of atomic oscillation, is measured at a fixed point in the moving frame. During that time interval, the moving frame will have moved with respect to the fixed frame, and therefore the times t_2 and t_1 will be measured at different positions in the stationary frame. From Eqs. (1-1),

$$T_0 = \left(t_2 - t_1 - \frac{vx_2}{c^2} + \frac{vx_1}{c^2}\right),$$

from which it follows that

$$T = \gamma T_0, \tag{1-3}$$

where $(x_2 - x_1)/T = v$. From Eq. (1-3), it follows that the interval between events is measured as shorter by the stationary observer than by an observer moving with the oscillator.

Now suppose further that the first pulse of radiation is emitted from time and position (t_0, x_0, y_0, z_0) of the coincident coordinate systems, and that at a time τ later a second pulse is emitted from the same position in C. The pulses are received at the coordinates (t', x', y', z'). Then,

$$\tau = \{(t' - t_0)^2 - (c^2)^{-1}[(x' - x_0)^2 + (y' - y_0)^2 + (z' - z_0)^2]\}^{1/2} \tag{1-4}$$

A property of the Lorentz transformations is that τ is independent of the speed of the frame in which it is measured, i.e., that it is invariant. Hence it is a scalar, and it is named the proper time. It has the dimensions of time, and it reduces to the time interval between pulses in that frame for which they were emitted from the same position in space.

If $i = (\sqrt{-1})$, the four coordinates (ict, x, y, z) are the components of a vector in four dimensions, i.e., a four-vector, since the sum of the squares of the components is the same in all reference frames. The symbol X_α will be used to indicate any one of the components of the positional four-vector:

$$X_\alpha = ict, x, y, z, \tag{1-5}$$

where α has integer values ranging between zero and three. By convention, any Greek index appearing twice in a single term is summed over all four indices, and any Greek index appearing once in a single term indicates direction.

Now a four-vector in which three of the terms asymptotically approach the velocity, for $v \ll c$, is

$$U_\alpha = \frac{dX_\alpha}{d\tau} = (i\gamma c, \gamma \mathbf{v}), \tag{1-6}$$

where $\mathbf{v}$ is the three-dimensional velocity. Since X_α is a four-vector, and τ is the same in all inertial frames, the sum of the squares of the components of U_α is the same in all reference frames. Similarly, one can define higher-order differential terms whose magnitude is the same in all reference frames; an example is the second-order term

$$W_\alpha = \frac{d^2 X_\alpha}{d\tau^2}. \tag{1-7}$$

In the limit of small velocities, at least, W_α is the acceleration. Substituting the values from Eqs. (1-4) and (1-5) into Eq. (1-6) gives

$$U_\alpha U_\alpha = -c^2, \tag{1-8}$$

which confirms that the magnitude of U is a constant for all frames. Differentiating Eq. (1-8) with respect to τ shows that

$$U_\alpha W_\alpha = 0, \tag{1-9}$$

i.e., that the velocity and acceleration vectors are perpendicular in four-space.

If motion at velocity $\mathbf{u}'$ in the moving frame is observed from both the stationary frame and the moving one, it is seen to be at different angles from the velocity axis. The speed and the angle of motion in the moving frame can be found by using Eqs. (1-1):

$$dx = \gamma(v\,dt' + dx'), \qquad dt = \gamma\left(dt' + \frac{v\,dx'}{c^2}\right),$$

$$dy = dy', \qquad dz = dz'.$$

So the differentials are

$$u_x = \frac{dx}{dt} = \frac{v + (dx'/dt')}{1 + (v\,dx'/c^2\,dt')} = \frac{v + u_x'}{1 + (vu_x/c^2)},$$
$$u_y = \frac{u_y'}{1 + (vu_x'/c^2)}, \qquad u_z = \frac{u_z'}{1 + (vu_x'/c^2)}. \tag{1-10}$$

Equations (1-10) are of particular interest when the moving object is light, or $u = u' = c$. Let α and α' be the angles between the radiation and the X and X' axes, respectively. By Eqs. (1-10), the tangent to the angle α, for velocity in the XY plane, is

$$\tan\alpha = \frac{u_y}{u_x} = \frac{u_y'}{\gamma(u_x' + v)}$$

or

$$\tan\alpha = \frac{\sin\alpha'}{\gamma(\cos\alpha' + \beta)}, \tag{1-11}$$

where $\beta = v/c$. Equation (1-11) is readily solved to give the differential relationship

$$d\alpha = \gamma(1 - \beta\cos\alpha)\,d\alpha'. \tag{1-12}$$

The difference between angles α and α' is the aberration of light. Light from the fixed stars appears to come from an angle that depends upon the velocity of the earth.

Equations (1-1) and (1-5) can be written as

$$X_\alpha' = C_{\alpha\beta}X_\beta, \tag{1-13}$$

where the primed frame is moving in the direction of increasing x with speed v. The constants $C_{\alpha\beta}$ are $C_{00} = C_{11} = \gamma$, $C_{10} = -C_{01} = i\beta\gamma$, $C_{22} = C_{33} = 1$, and all other C's equal 0. Since four-vectors are descriptive of physical phenomena, and since physical phenomena cannot be dependent upon how one chooses to describe them, all four-vectors transform as in Eqs. (1-13). For example, the four-velocities in the two frames are

$$U_\alpha' = C_{\alpha\beta}U_\beta. \tag{1-14}$$

Exercises

1. A rigid rod of length L_0 is at rest in a given frame of reference. It makes an angle α with respect to the velocity axis of a moving coordinate system. Show that the apparent length and orientation angle of the rod, as seen in the moving system, are

$$L = L_0\left(\frac{\cos^2\alpha}{\gamma^2} + \sin^2\alpha\right)^{1/2},$$

$$\tan\alpha' = \gamma\tan\alpha.$$

2. Consider an atom to be radiating at frequency ω_0. An observer moves at normalized speed β and at angle α with respect to the radiation emitted by the atom. The frequency shift is the Doppler shift. Show that the frequency shift is

$$\omega = \gamma\omega_0(1 + \beta\cos\alpha).$$

3. A space capsule is flying at 1000 m/sec. How much shorter will it appear to be to an observer on the ground? If the pilot and the ground observer synchronize watches, how long will it be before the pilot's watch is 1 μsec behind that of the ground observer?

4. Use the definition of proper time and X_α to verify the correctness of Eq. (1-6).

5. Verify that Eq. (1-8) is correct.

References

Aharoni, J., "The Special Theory of Relativity," 2nd ed., pp. 1–49. Oxford Univ. Press (Clarendon), London and New York, 1965.

Einstein, Albert, "The Meaning of Relativity," pp. 29–39. Princeton Univ. Press, Princeton, New Jersey, 1953.

Leighton, Robert B., "Principles of Modern Physics," pp. 1–40. McGraw-Hill, New York, 1959.

Lindsay, R. B., and H. Margenau, "Foundations of Physics," pp. 330–342. Dover, New York, 1957.

Panofsky, W. K. H., and M. Phillips, "Classical Electricity and Magnetism," 2nd ed., pp. 286–304. Addison–Wesley, Reading, Massachusetts, 1962.

Pauli, W., "Theory of Relativity," pp. 9–20. Pergamon Press, Oxford, 1958.

Sommerfeld, A., "Electrodynamics," pp. 223–235. Academic Press, New York, 1952.

2

ELECTROMAGNETIC TENSORS

If the Lorentz transformation equations are correct, then transformation must leave correctly stated physical laws unchanged in form. Mathematically, the laws are said to be Lorentz covariant. Therefore, if the description of an event as calculated by a certain law does not transform without change under a Lorentz transformation, then there is something wrong with either the law or the transformation. Since the Lorentz transformations are well supported by the cosmological principle, and since they have withstood experimental tests over wide ranges of distances, times, and velocities, we take them here to be correct for nonaccelerating, i.e., inertial, systems. Therefore, we must seek a set of quantitative statements of physical laws which remain covariant under Lorentz transformations and which simplify to the proper known forms for small velocities. There are two ways to construct such a set. The first is to begin with a known law in the limit of low velocity and to determine how it must be changed to become Lorentz covariant. The second is to construct Lorentz covariant tensors and extrapolate them to the low-velocity limit to connect them with known laws. Both methods will be used here.

Let us begin by postulating a scalar invariant quantity in addition to c and τ: the electric charge q. Since the physical extent of an electric charge is often much less than any macroscopic dimension of interest, and since the number of charges is very large, it is often convenient to work not with individual charges, but with *charge density*: the product of the number of charges per unit volume and the magnitude of the unit charge. Even though the apparent volume of a body depends upon its relative speed, the number of charges it contains does not; so that number will represent a scalar invariant quantity. The charge density ρ_0 is a continuous function of both time and position in any reference system; i.e., $\rho_0 = \rho_0(t, x, y, z)$.

The four-current density is defined to be

$$J_\alpha = \rho_0 U_\alpha \tag{2-1}$$

and is, of course, a four-vector. If electric charges are neither created nor destroyed, then

$$\frac{\partial J_\alpha}{\partial X_\alpha} = 0. \tag{2-2}$$

Many other four-tensors can be constructed on the basis of the current density. Let us define a series of them, examine their behavior at low velocities, and compare the findings with the results of applying physical laws deduced from experiments. A useful starting point is the differential equation

$$\frac{\partial^2 A_\alpha}{\partial X_\beta\, \partial X_\beta} = -\mu J_\alpha, \tag{2-3}$$

where μ is a constant whose value determines the system of units. Equation (2-3) is the defining equation for A_α. The result of combining Eqs. (2-2) and (2-3) is

$$\frac{\partial^2}{\partial X_\beta\, \partial X_\beta} \frac{\partial A_\alpha}{\partial X_\alpha} = 0.$$

Hence, $\partial A_\alpha / \partial X_\alpha$ has no sources anywhere in space and, therefore, must be constant. We put that constant equal to zero:

$$\frac{\partial A_\alpha}{\partial X_\alpha} = 0. \tag{2-4}$$

Other four-tensors can be constructed by using A_α. Of particular interest is the antisymmetric second-rank tensor

$$f_{\alpha\beta} = \frac{\partial A_\beta}{\partial X_\alpha} - \frac{\partial A_\alpha}{\partial X_\beta}. \tag{2-5}$$

Differentiating $f_{\alpha\beta}$ with respect to X_β and substituting into Eqs. (2-3) and (2-4) yields

$$\frac{\partial f_{\alpha\beta}}{\partial X_\beta} = \mu J_\alpha\,,$$

$$\frac{\partial f_{\nu\sigma}}{\partial X_\alpha} + \frac{\partial f_{\sigma\alpha}}{\partial X_\nu} + \frac{\partial f_{\alpha\nu}}{\partial X_\sigma} = 0. \tag{2-6}$$

Another four-vector $F_\alpha{}^{\mathrm{v}}$ can also be constructed:

$$F_\alpha{}^{\mathrm{v}} = f_{\alpha\beta} J_\beta\,. \tag{2-7}$$

By using the four-vectors as described, it can easily be verified that

$$F_\alpha{}^{\mathrm{v}} U_\alpha = 0. \tag{2-8}$$

A comparison of Eqs. (1-9) and (2-8) shows that both vectors $F_\alpha{}^{\mathrm{v}}$ and W_α are perpendicular to the four-velocity.

Still another useful four-tensor can be constructed by using the tensor $f_{\alpha\beta}$:

$$T_{\alpha\beta} = \mu^{-1}\left(f_{\alpha\lambda} f_{\lambda\beta} + \frac{\delta_{\alpha\beta}}{4} f_{\lambda\sigma} f_{\lambda\sigma}\right), \tag{2-9}$$

where $\delta_{\alpha\beta} = 1$ if $\alpha = \beta$, and is 0 otherwise. By algebraic manipulation, using Eqs. (2-5), (2-6), (2-7), and (2-9), it can be verified that

$$F_\alpha{}^{\mathrm{v}} = \frac{\partial T_{\alpha\beta}}{\partial X_\beta}. \tag{2-10}$$

This is a partial differential equation. If it is expressed in integral form, with the divergence theorem used for the last equality, it becomes

$$\int F_\alpha{}^{\mathrm{v}}\, dV^4 = \int \frac{\partial T_{\alpha\beta}}{\partial X_\beta}\, dV^4 = \int T_{\alpha\beta}\, dS_\beta\,, \tag{2-11}$$

where the four-volume $dV^4 = ic\ dt\ dx\ dy\ dz$ and the four-dimensional surfaces dS_β are perpendicular to the four-velocity U_β.

This completes the list of tensors required for a study of electrodynamics. It remains to define other, related vectors that are more convenient to work with in three dimensions, and then to relate those vectors to interactions of electric charges, i.e., to reconstruct the laws of classical electrodynamics.

The two vectors that are particularly useful in three-dimensional space are the electric and magnetic field vectors **E** and **B**, defined as

$$E_\alpha = ic\left(\frac{\partial A_0}{\partial X_\alpha} - \frac{\partial A_\alpha}{\partial X_0}\right),$$
$$B_\alpha = \frac{\partial A_\gamma}{\partial X_\beta} - \frac{\partial A_\beta}{\partial X_\gamma}, \tag{2-12}$$

where $\alpha, \beta, \gamma = 1, 2,$ or 3. When Eqs. (2-5), (2-6), and (2-12) are combined, the result is

$$\nabla \times \mathbf{E} + \frac{\partial \mathbf{B}}{\partial t} = 0, \qquad \nabla \cdot \mathbf{E} = \mu c^2 \rho,$$
$$\nabla \times \mathbf{B} - (c^2)^{-1}\frac{\partial \mathbf{E}}{\partial t} = \mu \mathbf{J}, \qquad \nabla \cdot \mathbf{B} = 0. \tag{2-13}$$

These are the Maxwell equations. The four-vector $F_\alpha{}^{\mathrm{v}}$ can be rewritten in terms of the three-dimensional quantities by use of Eqs. (2-7) and (2-12); the result is

$$\mathbf{F}^{\mathrm{v}} = \mathbf{J} \times \mathbf{B} + \rho \mathbf{E},$$
$$-icF_0{}^{\mathrm{v}} = \mathbf{E} \cdot \mathbf{J}. \tag{2-14}$$

The symmetric four-tensor $T_{\alpha\beta}$ can be expressed in terms of **E** and **B** as

$$
\begin{aligned}
T_{00} &= \frac{\varepsilon}{2}\,\mathbf{E}\cdot\mathbf{E} + (2\mu)^{-1}\mathbf{B}\cdot\mathbf{B},\\
T_{10} &= \frac{-i}{\mu c}(E_yB_z - E_zB_y),\\
T_{20} &= \frac{-i}{\mu c}(E_zB_x - E_xB_z),\\
T_{30} &= \frac{-i}{\mu c}(E_xB_y - E_yB_x),\\
T_{11} &= \frac{\varepsilon}{2}(E_x^{\,2} - E_y^{\,2} - E_z^{\,2}) + (2\mu)^{-1}(B_x^{\,2} - B_y^{\,2} - B_z^{\,2}),\\
T_{12} &= \varepsilon E_xE_y + \mu^{-1}B_xB_y,\\
T_{13} &= \varepsilon E_zE_x + \mu^{-1}B_zB_x,\\
T_{22} &= \frac{\varepsilon}{2}(E_y^{\,2} - E_z^{\,2} - E_x^{\,2}) + (2\mu)^{-1}(B_y^{\,2} - B_z^{\,2} - B_x^{\,2}),\\
T_{23} &= \varepsilon E_yE_z + \mu^{-1}B_yB_z,\\
T_{33} &= \frac{\varepsilon}{2}(E_z^{\,2} - E_x^{\,2} - E_y^{\,2}) + (2\mu)^{-1}(B_z^{\,2} - B_x^{\,2} - B_y^{\,2})
\end{aligned}
\tag{2-15}
$$

where

$$\varepsilon = (\mu c^2)^{-1}.$$

The tensor $T_{\alpha\beta}$ can be conveniently handled by breaking it into three parts, of which one contains only the spatial components, one contains both time and spatial components, and the third contains only the time components, i.e.,

$$
T_{\alpha\beta} = \left[\begin{array}{c|c} W & -\frac{i}{c}\mathbf{N} \\ \hline -\frac{i}{c}\mathbf{N} & \sigma_{ij} \end{array}\right],
$$

where **N** is the Poynting vector, given by

$$\mathbf{N} = \mu^{-1}(\mathbf{E}\times\mathbf{B}), \tag{2-16}$$

W is a three-dimensional scalar given by

$$W = \tfrac{1}{2}(\varepsilon \mathbf{E} \cdot \mathbf{E} + \mu^{-1} \mathbf{B} \cdot \mathbf{B}), \tag{2-17}$$

and σ_{ij} is a second-rank, symmetric three-tensor given by

$$\sigma_{ij} = \varepsilon(E_i E_j - \tfrac{1}{2}\delta_{ij} E_k E_k) + \mu^{-1}(B_i B_j - \tfrac{1}{2}\delta_{ij} B_k B_k). \tag{2-18}$$

This completes the writing of the four tensors in terms of the three-dimensional field quantities.

Four-tensors, just like four-vectors, transform by means of Lorentz transformations. If they did not, they would form differently in the different inertial frames and would not all be equivalent. For example, the tensor $f_{\alpha\beta}$ transforms as

$$f'_{\alpha\beta} = C_{\alpha\delta} C_{\beta\eta} f_{\delta\eta}\,. \tag{2.19}$$

$f_{\alpha\beta}$ is a second-rank four-tensor. The rank of a tensor is equal to the number of indices. A tensor of rank zero, τ for example, is independent of the speed of the frame. A four-vector is a tensor of rank one, etc. A tensor of rank r consists of an array of D^r terms, where D is the number of dimensions.

Exercises

1. Find the Lorentz transformation equations for the electromagnetic potentials.

Ans.: $\Phi' = \gamma(\Phi - \beta c A_x)$, $\quad A_x' = \gamma\left(A_x - \dfrac{\beta\Phi}{c}\right)$,

$A_y' = A_y$, $\quad A_z' = A_z$.

2. Find the Lorentz transformation equations for the current densities.

Ans.: $\rho' = \gamma\left(\rho - \dfrac{\beta J_x}{c}\right)$, $\quad J_x' = \gamma(J_x - \beta c \rho)$,

$J_y' = J_y$, $\quad J_z' = J_z$.

3. Find the Lorentz transformation equations for the electric and magnetic fields by using the transformation on $f_{\alpha\beta}$.

Ans.: $B_x' = B_x$, $\qquad E_x' = E_x$,

$$B_y' = \gamma\left(B_y + \frac{\beta E_z}{c}\right), \qquad E_y' = \gamma(E_y - \beta c B_z),$$

$$B_z' = \gamma\left(B_z - \frac{\beta E_y}{c}\right), \qquad E_z' = \gamma(E_z + \beta c B_y).$$

These transformation equations are often written in a form more convenient for working with three-dimensional vectors:

$$\mathbf{B}_{\|}' = \mathbf{B}_{\|}, \qquad \mathbf{B}_{\perp}' = \gamma\left(\mathbf{B}_{-} - \frac{\mathbf{v} \times \mathbf{E}}{c^2}\right),$$

$$\mathbf{E}_{\|}' = \mathbf{E}_{\|}, \qquad \mathbf{E}_{\perp}' = \gamma(\mathbf{E}_{\perp} + \mathbf{v} \times \mathbf{B}).$$

4. Show that the equations of motion of a charged particle of charge q and mass m, moving in an electric and magnetic field, are

$$m \frac{d^2 X_\alpha}{d\tau^2} = q f_{\alpha\beta} U_\beta . \qquad (2\text{-}20)$$

5. Show that the following quantities are four-dimensional tensors:

(a) $c^2\mathbf{A} \cdot \mathbf{A} - \Phi^2$, $\qquad$ (b) $\mathbf{A} \cdot \mathbf{J} - \rho\Phi$,

(c) $\varepsilon\mathbf{E} \cdot \mathbf{E} - \mu^{-1}\mathbf{B} \cdot \mathbf{B}$, $\quad$ (d) $\mathbf{J} \cdot \mathbf{J} - \rho^2 c^2$.

6. Show that, if **E** and **B** are perpendicular in one Lorentz frame, they are perpendicular in all Lorentz frames.

7. Show that, if **E** is perpendicular to **B** and if $E \neq cB$, then there is a frame in which the field is either purely electric or purely magnetic.

8. Combine Eqs. (2-5), (2-7), and (2-9) to obtain Eq. (2-10).

REFERENCES

Aharoni, J., "The Special Theory of Relativity," 2nd ed., pp. 89–124. Oxford Univ. Press (Clarendon), London and New York, 1965.

Jackson, J. D., "Classical Electrodynamics," pp. 374–386. Wiley, New York, 1962.

Leighton, Robert B., "Principles of Modern Physics," pp. 40–48. McGraw-Hill, New York, 1959.

Panofsky, W. K. H., and M. Phillips, "Classical Electricity and Magnetism," 2nd ed., pp. 324–336. Addison-Wesley, Reading, Mass., 1962.

Sommerfeld, A., "Electrodynamics," pp. 212–222, 237–249. Academic Press, New York, 1952.

3

EVALUATION OF THE FOUR-POTENTIAL

The quantities **E** and **B** can be readily calculated if the potential function A_α, defined by Eq. (2-3), is known. But that quantity is known only indirectly, through its relationship to the current density J_α. The problem, then, is to evaluate A_α by integrating Eq. (2-3). For this purpose, it is convenient to use the Green's function $G(X_\alpha , X_\alpha')$, where G is by definition that function which satisfies the differential equation

$$\frac{\partial^2 G(X_\alpha, X_\alpha')}{\partial X_\beta \, \partial X_\beta} = - [\delta(X_\alpha - X_\alpha')]^4, \tag{3-1}$$

where $\delta(X_\alpha - X_\alpha')$ is the Dirac delta function. The integral of a delta function is given by

$$\int f(x) \, \delta(x - x_0) \, dx = f(x_0) \tag{3-2}$$

if the range of integration includes x_0, and is zero otherwise. Since $\int \delta[g(x) - g(x_0)] \, dg = \int \delta(x - x_0) \, dx$, and since $df = f' \, dx$,

$$\int f(x) \, \delta[g(x) - g(x_0)] \, dx = \frac{g(x_0)}{f'(x_0)} \tag{3-3}$$

if, as before, the range of integration includes x_0, and is zero otherwise.

To proceed with the calculation of the potential, it is useful to construct the quadruple integral

$$\int\int\int\int\left(A_\alpha \frac{\partial^2 G}{\partial X_\beta\, \partial X_\beta} - G \frac{\partial^2 A_\alpha}{\partial X_\beta\, \partial X_\beta}\right) dV^4 , \tag{3-4}$$

where $dV^4 = ic\ dt\ dx\ dy\ dz$. Equation (3-4) can be rewritten as a sum of surface integrals in four-space, one of which is

$$\int\int\int_{-\infty}^{\infty}\left(A_\alpha \frac{\partial G}{\partial t} - G \frac{\partial A_\alpha}{\partial t}\right) dx\ dy\ dz.$$

All integrals of this form are assumed to be equal to zero, i.e., it is assumed that all results vanish in the limit as each coordinate becomes infinite. Therefore, the integral of Eq. (3-4) must be zero; so combining it with Eqs. (2-3), (3-1), and (3-2) gives

$$A_\alpha(X_\gamma) = \mu \int\int\int\int J_\alpha(X_\gamma{}')G(X_\gamma{}', X_\gamma)\, dV'^4 . \tag{3-5}$$

Equation (3-5) is in the desired form, since, if G and J_α are known, A_α can be calculated at all points.

The problem now is to find G. For this purpose, it is useful to bring in the Fourier transform relations:

$$F(X_\gamma) = \left[\frac{1}{2\pi}\right]^2 \int\int\int\int_{-\infty}^{\infty} H(K_\gamma) \exp(-iX_\gamma K_\gamma)\, dV_K{}^4 ,$$
$$H(K_\gamma) = \left[\frac{1}{2\pi}\right]^2 \int\int\int\int_{-\infty}^{\infty} F(X_\gamma) \exp(iX_\gamma K_\gamma)\, dV^4 . \tag{3-6}$$

Substituting $\delta(X_\gamma - X_\gamma{}')$ for $F(X_\gamma)$ in these equations gives

$$[\delta(X_\gamma - X_\gamma{}')]^4 = [(2\pi)^4]^{-1} \int\int\int\int \exp[-iK_\gamma(X_\gamma - X_\gamma{}']\, dV_K{}^4, \tag{3-7}$$

where the K_γ are the conjugate variables to the X_γ, and $dV_K{}^4 = dK_0\, dK_x\, dK_y\, dK_z$. Now G can also be expressed as an integral over the K's, thus:

$$G(X_\gamma , X_\gamma{}') = \int\int\int\int dV_K{}^4\, g(K_\gamma) \exp[-iK_\gamma(X_\gamma - X_\gamma{}')]. \tag{3-8}$$

Combining Eqs. (3-1), (3-7), and (3-8) gives

$$g(K_\gamma) = [(2\pi)^4 K_\alpha K_\alpha]^{-1}.$$

Since the four-vector X_α is a complex quantity, it appears that K_α should also be complex. It is convenient to rewrite the zeroth component of K_α, explicitly setting forth the complex form as $K_0 = i\omega/c$. This form emphasizes the singularity in the foregoing expression for $g(K_\gamma)$:

$$g(K_\gamma) = \frac{-c^2}{(2\pi)^4[k^2 - (\omega^2/c^2)]}, \tag{3-9}$$

where k^2 is the sum of the squares of each of the components in three dimensions.

The integral of Eq. (3-8), when Eq. (3-9) is substituted, is a Cauchy integral, which can be evaluated by using the integral identity

$$\oint \frac{f(z)\,dz}{z - z_0} = 2\pi i f(z_0) \tag{3-10}$$

and writing the integral for G as

$$G(X_\gamma, X_\gamma') = \frac{c}{i(2\pi)^4} \int\int\int dk^3 \int \frac{d\omega \exp[-iK_\alpha(X_\alpha - X_\alpha')]}{(\omega - ck + ip)(\omega + ck + ip)}, \tag{3-11}$$

where $dk^3 = dk_x\, dk_y\, dk_z$, and where, for convenience in evaluating the integral, p is introduced as a small, real, positive number. The problem is, given an electric charge at the time and position $(t', \mathbf{r}')$, to find the function $G(X_\gamma, X_\gamma')$ and thus the potential $A_\alpha(t, \mathbf{r})$. All fields are zero before the charge is introduced; i.e., if $t'' = t - t'$, all fields are zero for $t'' < 0$. The integral can now be evaluated by integrating first along the real axis and then back around an infinite, complex ω path. For $t'' < 0$, the return path encompasses the lower half-plane, and no poles are enclosed. For $t'' > 0$, the return path is around the upper half of the complex plane, and two poles are enclosed. The result is

$$G(X_\gamma, X_\gamma') = \frac{i}{(2\pi)^3} \int\int\int \frac{dk^3}{k} \sin ckt'' \exp(-i\mathbf{k} \cdot \mathbf{R}), \tag{3-12}$$

where **R** is the distance between the source point **r**′ and the field point **r**. Equation (3-12) is a triple integral over the three-vector **k**. The integrals can be evaluated by going to k-space, where **R** is chosen to be along the z axis:

$$dk^3 = k^2\,dk \sin\theta\,d\theta\,d\phi,$$

$$\mathbf{k}\cdot\mathbf{R} = kR\cos\theta,$$

so that the integral becomes

$$G(X_\gamma, X_\gamma') = \frac{2i}{(2\pi)^2}\int_0^\infty dk \sin ckt'' \sin kR. \tag{3-13}$$

Since the integral is an even function of k, it can be replaced by

$$G(X_\gamma, X_\gamma') = \frac{i}{8\pi^2 R}\int_{-\infty}^{\infty} dk$$

$$\cdot\{\exp[i(\omega t'' - kR)] - \exp[-i(\omega t'' + kR)]\}. \tag{3.14}$$

Each term in Eq. (3-14) is a delta function. Since the second one is evaluated at $t'' < 0$, all results from it are zero when substituted into Eq (3-5). We therefore ignore it. The first term gives

$$G(X_\gamma, X_\gamma') = (4\pi Ric)^{-1}\,\delta\left(t'' - \frac{R}{c}\right), \tag{3-15}$$

since

$$\delta(R - ct'') = -c^{-1}\,\delta\left(t'' - \frac{R}{c}\right).$$

Now that G is known, we can substitute it back into Eq. (3-5) to determine A_α:

$$A_\alpha(X_\gamma) = \frac{\mu}{4\pi ic}\int\int\int\int dV'^4\,\frac{J_\alpha(X_\gamma')}{R}\,\delta\left(t'' - \frac{R}{C}\right). \tag{3-16}$$

This equation gives the potential $A_\alpha(X_\gamma)$ at each position in four-space as a result of the current density at each position in four-space. It can be rewritten as

$$A_\alpha(X_\gamma) = \frac{\mu}{4\pi}\int\int\int dV'^3 \int dt'\,\frac{J_\alpha(X_\gamma')}{R(X_\gamma, X_\gamma')}\,\delta\left(t' + \frac{R}{c} - t\right), \tag{3-17}$$

where dV'^3 is a differential volume in three-space. The combination of Eqs. (3-3) and (3-17) shows that

$$A_\alpha(X_\gamma) = \frac{\mu}{4\pi} \int\int\int dV'^3 \frac{J_\alpha[t - (R/c'), x_\gamma']}{(1 - \hat{R} \cdot \boldsymbol{\beta}) R(X_\gamma, X_\gamma')}, \qquad (3\text{-}18)$$

since $[\partial R(t', x_\gamma')/\partial t']_{x\gamma} = -c\hat{R} \cdot \boldsymbol{\beta}$. The lower-case letters indicate a variable in three dimensions. Equation (3-18) describes the four-potential in its most useful form.

REFERENCE

Jackson, J. D., "Classical Electrodynamics," pp. 183–186. Wiley, New York, 1962.

4
ELECTRODYNAMICS

In this chapter, we first combine the definitive differential equation (2-3) with the Green's function development of Chapter 3 to obtain an expression for A_α. Then, having found A_α, we use the differential relationships developed in Chapter 2 to obtain explicit forms for the field vectors and the force laws.

An electric charge is said to be a point charge if all variations within it occur over distances much less than other distances under consideration and if it is not subject to deformation. Classical physics is concerned primarily with the behavior and properties of point charges. By Eq. (3-17), the potential of a point charge of magnitude q is

$$A_\alpha(t, \mathbf{r}) = \frac{\mu q}{4\pi\gamma}\int dt' \, \frac{U_\alpha(t')}{R(t')} \, \delta\left(t' + \frac{R(t')}{c} - t\right). \tag{4-1}$$

Substituting Eq. (3-3) into Eq. (4-1) gives

$$A_\alpha(t, \mathbf{r}) = \frac{\mu q}{4\pi\gamma} \frac{U_\alpha[\mathrm{t} - (R/c)]}{KR}, \tag{4-2}$$

where

$$K = 1 - \hat{R} \cdot \boldsymbol{\beta}. \tag{4-3}$$

$\hat{R}$ is a unit vector directed from the source point to the field point, and $\boldsymbol{\beta}$ is the velocity of the charge as measured in the frame of the field point at the time $[t - (R/c)]$.

The four components of Eq. (4-2) are called the Liénard–Wiechert potentials. Their physical significance is most easily seen by going to the low-velocity limit and working in three-dimensional space. Separating the last three components of A_α from the first gives

$$A(t, \mathbf{r}) = \frac{\mu q}{4\pi RK} \mathbf{v}(\mathrm{t} - R/c),$$
$$- icA_0 = \Phi(t, \mathbf{r}) = \frac{q}{4\pi\varepsilon KR}, \tag{4-4}$$

which serves to define the potential, $\Phi(t, \mathbf{r})$. Now the problem is to calculate the electric and magnetic fields about a charge in motion, that is, the fields at time and position $(t, \mathbf{r})$ generated by a charge that is at time and position $(t', \mathbf{r}')$ and that has velocity $\boldsymbol{\beta}(t')$. The normalized velocity and acceleration of the charge are

$$\boldsymbol{\beta}(t') = \left.\frac{-\partial\mathbf{R}(t')}{c\partial t'}\right|_{\mathbf{r}},$$
$$\dot{\boldsymbol{\beta}}(t') = \left.\frac{-\partial^2 R(t')}{c\partial t'^2}\right|_{\mathbf{r}}.$$

The distance between the source and the field points is

$$R(\mathbf{r}, \mathbf{r}') = c(t - t').$$

Therefore,

$$\left.\frac{\partial R}{\partial t}\right|_{\mathbf{r}} = \frac{dR}{dt'}, \qquad \left.\frac{\partial t'}{\partial t}\right|_{\mathbf{r}} = -c(\boldsymbol{\beta} \cdot \hat{R}) \left.\frac{\partial t'}{\partial t}\right|_{\mathbf{r}} = c\left(1 - \left.\frac{\partial t'}{\partial t}\right|_{\mathbf{r}}\right),$$

and finally

$$\left.\frac{\partial t'}{\partial t}\right|_{\mathbf{r}} = (1 - \hat{R} \cdot \boldsymbol{\beta})^{-1} = K^{-1}. \tag{4-5}$$

In a similar way, the del operator operates on t' to give

$$\nabla t' = -\frac{\hat{R}}{cK}.$$

Let us use Eqs. (2-13) to operate on the potentials of Eqs. (4-4); this gives, for the field intensities in the vicinity of the charge,

$$\mathbf{E} = \frac{q(\mathbf{R} - R\boldsymbol{\beta})}{4\pi\varepsilon R^3 K^3 \gamma^2} + \frac{\eta q}{4\pi R^3 K^3} \mathbf{R} \times [(\mathbf{R} - R\boldsymbol{\beta}) \times \dot{\boldsymbol{\beta}}], \tag{4-6}$$

$$\mathbf{B} = c^{-1} (\hat{R} \times \mathbf{E}), \tag{4-7}$$

where $\eta = (\mu/\varepsilon)^{1/2}$. Equations (4-6) and (4-7) describe the electric and magnetic fields about a rigid point charge in motion. The distance **R** is measured from the retarded position of the charge to the source position.

The theorems of classical mechanics, i.e., Newton's laws, follow from combining Eqs. (4-6) and (4-7) with the integral equation (2-11). Classical mechanics is applicable in the limit of speeds low enough that $\beta \ll 1$ and $K \approx 1$. For accelerations low enough that the first term in the brackets predominates, the fields become

$$\mathbf{E} = \frac{q\hat{R}}{4\pi\varepsilon R^2},$$

$$\mathbf{B} = \frac{\eta q}{4\pi R^2} (\boldsymbol{\beta} \times \hat{R}). \tag{4-8}$$

The expression for **E** is Coulomb's law, and the expression for **B** is Ampere's law. Now let us evaluate the surface integrals of Eq. (2-11) over each surface. The three spatial variables have the surface values of $\pm\infty$, and X_0 has the surface values of $-i\infty$ and ict. Substituting **E** and **B** from Eqs. (4-8) into the integrand on each of the eight surfaces gives zero for seven of them, but, for the eighth, given by $X_0 = ict$, the integral is nonzero. The shape of a rigid object, however, as seen from a particle in motion with respect to it, is velocity-dependent. And the volume-to-surface transformation of Eq. (2-11) is correct only when the surface integrals range over surfaces perpendicular to the coordinates.

A charge q, stationary in the primed frame, is moving in the positive x direction with respect to an observer in the unprimed frame. In the frame of the charge, we first multiply Eq. (2-10) by $C'_{\sigma\alpha}$ and then integrate over the primed coordinates to obtain

$$\int C'_{\sigma\alpha} F_{\sigma}^{\mathrm{v}\prime} \, dV'^4 = \int C'_{\sigma\alpha} \frac{\partial T'_{\alpha\beta}}{\partial X'_{\beta}} \, dV'^4. \tag{4-9}$$

From the left side of Eq. (4-9), we obtain

$$\begin{aligned} \int C'_{\sigma\alpha} F_{\alpha}^{\mathrm{v}\prime} \, dV'^4 &= \int F_{\sigma}{}^{\mathrm{v}} \, dV'^4 = \int F_{\sigma}{}^{\mathrm{v}} \, dV^4 \, | C_{\alpha\beta} | \\ &= \int F_{\sigma}{}^{\mathrm{v}} \, dV^4, \end{aligned} \tag{4-10}$$

using the usual rules for changing variables of an integrand and noting that $|C_{\alpha\beta}| = 1$. Working with the right side of Eq. (4-9), we obtain

$$\begin{aligned} \int C'_{\sigma\alpha} \frac{\partial T'_{\sigma\beta}}{\partial X_{\beta}'} \, dV'^4 &= \int \frac{\partial}{\partial X_{\beta}'} (C'_{\sigma\alpha} T'_{\alpha\beta}) \, dV'^4 \\ &= \oint C'_{\sigma\alpha} T'_{\alpha\beta} \, dS'^4 \\ &= \int C'_{\sigma\alpha} T'_{\alpha 0} \, dV'^3, \end{aligned} \tag{4-11}$$

where the four-volume integral changed to a four-surface integral, and all surface integrals save that at positive time t' were put equal to zero.

To work with the final integral of Eq. (4-11), we first note that

$$C'_{\lambda\beta} T_{\sigma\lambda} = C'_{\lambda\beta} C'_{\sigma\alpha} C'_{\lambda\gamma} T'_{\alpha\gamma}$$

and, working with the transformation matrices, that

$$C'_{\lambda\beta} C'_{\lambda\gamma} = \delta_{\beta\gamma} \, ; \tag{4-12}$$

hence,

$$C'_{\lambda\beta} T'_{\sigma\lambda} = C'_{\sigma\alpha} T'_{\alpha\beta} \, . \tag{4-13}$$

The combination of Eqs. (4-9), (4-10), (4-11), and (4-13) gives

$$\int F_{\sigma}{}^{\mathrm{v}} \, dV^4 = \int C_{0\alpha} T_{\sigma\alpha} \, dV^3 \tag{4-14}$$

after noting that $C'_{\lambda\beta} = C_{\beta\lambda}$. Comparison of the transformation matrix with the four-velocity shows that

$$C_{\alpha 0} = \frac{U_{\alpha}}{ic}, \tag{4-15}$$

and combining Eqs. (4-14) and (4-15) gives

$$\int F_\sigma{}^{v}\, dV^4 = (ic)^{-1} \int T_{\sigma\alpha} U_\alpha\, dV^3. \tag{4-16}$$

Differentiation of Eq. (4-16) by t completes the analysis to give

$$F_\sigma = \int F_\sigma{}^{v}\, dV^3 = -(c^2)^{-1} \frac{d}{dt} \int T_{\sigma\alpha} U_\alpha\, dV^3. \tag{4-17}$$

We now change the sign of Eq. (4-17) to represent reaction of the field on its source, rather than vice versa, and evaluate results of Eq. (4-17) for the case of small acceleration. Also from Eqs. (4-6) and (4-7) it follows that

$$\mathbf{B} = c^{-1} \boldsymbol{\beta} \times \mathbf{E}. \tag{4-18}$$

Putting Eq. (4-18) into (4-17) and putting $\sigma = 0$ gives

$$F_0 = \frac{i}{c} \int \mathbf{E} \cdot \mathbf{J}\, dV^3 = \frac{i}{c} \frac{d}{dt} \int dV^3 \gamma \left[\frac{\varepsilon}{2} E^2 - (2\mu)^{-1} B^2 \right], \tag{4-19}$$

whereas the nonzero components of $\alpha = j$ give

$$\begin{aligned} F_j &= \int dV^3 [\rho E_j + (\mathbf{J} \times \mathbf{B})_j] \\ &= (c^2)^{-1} \frac{d}{dt} \int \left\{ c\beta_j \gamma \left[\frac{\varepsilon}{2} E^2 - (2\mu)^{-1} B^2 \right] \right\} dV^3. \end{aligned} \tag{4-20}$$

Equation (4-20) can be more conveniently written as

$$F_j = \frac{d}{dt}(mv_j), \tag{4-21}$$

where

$$m = (2c^2)^{-1} \int \gamma \left(\varepsilon E^2 - \frac{B^2}{\mu} \right) dV^3. \tag{4-22}$$

Equation (4-21) is Newton's law. It can be put in four-dimensional form by defining momentum density to be the integrand of Eq. (4-17):

$$g_\sigma = (c^2)^{-1} T_{\sigma\alpha} U_\alpha, \tag{4-23}$$

and, using Eq. (4-17), the force law can be written as

$$F_\sigma = \frac{d}{dt} G_\sigma , \tag{4-24}$$

where

$$G_\sigma = (icm, m\mathbf{v}) \tag{4-25}$$

with mass given by Eq. (4-22). It should be kept in mind that Eqs. (4-21), (4-22), and (4-25) are to be used for finding the acceleration. Yet Eqs. (4-9) and (4-18) are true only when the acceleration is vanishingly small. Therefore, the preceding results will be invalid for large values of acceleration.

Returning to Eqs. (2-7) and (2-10) and the equation

$$\frac{\partial T_{\alpha\beta}}{\partial X_\beta} = f_{\alpha\beta} J_\beta , \tag{4-26}$$

we now evaluate Eq. (4-24) in terms of the field quantities. For j and k ranging from 1 to 3, the result is

$$\rho E_j + (\mathbf{J} \times \mathbf{B})_j = -(c^2)^{-1} \frac{\partial N_j}{\partial t} + \frac{\partial \sigma_{jk}}{\partial x_k} , \tag{4-27}$$

where σ_{jk} is defined in Eq. (2-18). In the static case, the first term on the right is zero; therefore, the second term is the static force per unit volume. The combination of Eqs. (4-24), (4-26), and (4-27) shows the static force to be

$$\int \sigma_{jk}\, dS_k = \int \frac{\partial \sigma_{jk}}{\partial x_k}\, dV = \frac{d}{dt} \int [\mathbf{g}_j + (c^2)^{-1} N_j]\, dV, \tag{4-28}$$

where $\mathbf{g}$ is the momentum density. It follows, then, that the expression

$$\mathbf{g}_{\text{em}} = (c^2)^{-1} \mathbf{N} \tag{4-29}$$

is the momentum density associated with the electromagnetic field.

The evaluation of Eq. (4-25) for the $\alpha = 0$ term in terms of three-dimensional fields gives

$$-\mathbf{E}\cdot\mathbf{J} = \frac{\partial}{\partial t}\left[\frac{\varepsilon}{2}\mathbf{E}\cdot\mathbf{E} + (2\mu)^{-1}\mathbf{B}\cdot\mathbf{B}\right] + \nabla\cdot\mathbf{N}. \tag{4-30}$$

Since **E** is the force on a unit positive charge, $\mathbf{E}\cdot\mathbf{J}$ is the power density applied to the current. By integrating over a finite volume and by changing the sign of the left side to give the power to the volume, we obtain

$$\int \mathbf{E}\cdot\mathbf{J}\,dV = -\int dV\frac{d}{dt}\left[\frac{\varepsilon}{2}\mathbf{E}\cdot\mathbf{E} + (2\mu)^{-1}\mathbf{B}\cdot\mathbf{B}\right] + \oint \mathbf{N}\cdot dS. \tag{4-31}$$

Since the left side is the power input, the expression

$$W = \frac{1}{2}\int dV\left[\frac{\varepsilon}{2}\mathbf{E}\cdot\mathbf{E} + (2\mu)^{-1}\mathbf{B}\cdot\mathbf{B}\right] \tag{4-32}$$

is the energy stored in the fields, and

$$P = \oint \mathbf{N}\cdot d\mathbf{S} \tag{4-33}$$

is the power escaping through the surface.

The surface integral in Eq. (4-31) is zero for nonaccelerating charges. Then a comparison of Eqs. (4-19), (4-22), (4-31), and (4-32) shows

$$m = (2c^2)^{-1}\int dV^3\left(\varepsilon E^2 + \frac{B^2}{\mu}\right), \tag{4-34}$$

i.e., energy is related to the mass by the simple relationship $W = mc^2$. From Exercise 5, Chapter 2, the term in parentheses in the integrand of Eq. (4-22) is the same in all Lorentz frames. Therefore, the mass increases in proportion to γ.

The fields many particle diameters away from the particle are described by Eq. (4-6), but the fields near the particles are dependent upon the structure of the particles, and so are not known. To apply the results of Eq. (4-34) to an electron, it has been pictured as a spherical shell of charge; this is the Lorentz model. For this model, since the mass m of an electron at zero speed is known, we can calculate the effective diameter from Eq. (4-34), as follows:

$$D_1 = \frac{\mu e^2}{4\pi m}, \tag{4-35}$$

which is numerically equal to 2.818×10^{-15}m.

Next we consider the case in which the second term in the brackets of Eq. (4-6) predominates over the first term, but the speed of the particle is still small enough so that $K \approx 1$,

$$\begin{aligned} \mathbf{E} &= \frac{\eta q}{4\pi R}[\hat{R} \times (\hat{R} \times \dot{\boldsymbol{\beta}})], \\ \mathbf{B} &= \frac{\mu q}{4\pi R}\dot{\boldsymbol{\beta}} \times \hat{R}, \end{aligned} \tag{4-36}$$

and the Poynting vector is

$$\mathbf{N} = \frac{\eta q^2 \hat{R}}{16\pi^2 R^2}(\hat{R} \times \dot{\boldsymbol{\beta}})^2 \tag{4-37}$$

at each instant. The net flux of the Poynting vector away from the charge is calculated by integrating over a sphere based upon the charge, which gives

$$\oint \mathbf{N} \cdot d\mathbf{S} = \frac{\eta q^2 \dot{\beta}^2}{6\pi} \tag{4-38}$$

for the radiated power.

We now turn to the electrostatic potential $\Phi(x_\alpha)$ produced by a fixed array of static charges, subject to the condition that the distance between an origin at any fixed point near the array and the field point R is much greater than the largest interarray distance. It follows from Eq. (4-4) and the definition of charge density that the potential is

$$\Phi(x_\alpha) = (4\pi\varepsilon)^{-1} \int \frac{\rho(x')\, dV'}{R(x_\alpha, x_\alpha')}, \tag{4-39}$$

where

$$\mathbf{r} = \mathbf{R} + \mathbf{r}'.$$

The distance R, expanded into a Taylor series, is

$$R^{-1} = r^{-1} + \frac{\partial}{\partial x_\alpha'}(R^{-1})\big|_{\mathbf{r}} x_\alpha' + \frac{1}{2!}\frac{\partial^2}{\partial x_\alpha' \, \partial x_\beta'}(R^{-1})\big|_{\mathbf{r}} x_\alpha' x_\beta' + \cdots, \tag{4-40}$$

and Eqs. (4-39) and (4-40) combine to give

$$\Phi(x_\gamma) = (4\pi\varepsilon)^{-1}\left[r^{-1}\int \rho(x_\gamma')\,dV' + \frac{\partial}{\partial x_\alpha'}(R^{-1})\big|_{\mathbf{r}}\int \rho(x_\gamma')x_\alpha'\,dV' + \frac{1}{2!}\frac{\partial^2}{\partial x_\alpha' \, \partial x_\beta'}(R^{-1})\big|_{\mathbf{r}}\int \rho(x_\gamma')x_\alpha' x_\beta'\,dV' + \cdots\right]. \tag{4-41}$$

Since each term in Eq. (4-41) is one order of magnitude less than the preceding one, the series converges rapidly. This approach is particularly useful when $\int \rho(x_\gamma')\,dV' = 0$, and for that case the potential is very nearly equal to

$$\Phi(x_\gamma) = \frac{r_\alpha}{4\pi\varepsilon r^3}\int \rho(x_\gamma')x_\alpha'\,dV'. \tag{4-42}$$

The integral of Eq. (4.42) is defined to be the dipolar moment $\mathbf{p}$, or

$$\mathbf{p} = \int \rho(x_\gamma')\mathbf{r}'\,dV'. \tag{4-43}$$

The combination of Eqs. (4-42) and (4-43) shows the potential to be

$$\Phi(x_\gamma) = (4\pi\varepsilon)^{-1}\frac{\mathbf{p}\cdot\mathbf{R}}{R^2}. \tag{4-44}$$

The electric field intensity follows from combining Eqs. (2-12), (4-4), and (4-42):

$$\mathbf{E} = \frac{p}{4\pi\varepsilon R^3}[2\hat{R}\cos\theta + \hat{\theta}\sin\theta] \tag{4-45}$$

for a Z-directed dipolar moment.

Next let us consider a continuous charge distribution arranged in the shape of a torus with radius a, but of negligible thickness, and of total charge q. The torus spins about an axis through its center with peripheral speed v. We wish to find the fields produced. The potentials are given by Eq. (4-2) for a point charge q, where R and U_α are evaluated at the retarded time, i.e., the time R/c earlier than the time of measurement of the field. Let us place the torus in the X–Y plane with its center at the origin. We evaluate the potentials in the X–Z plane. The differential charge density is

$$dq = q\frac{d\phi_t}{2\pi},$$

where $d\phi_t$ is measured in terms of time t in the fixed system. Since R and v are measured at retarded time t', we must either transform $d\phi_t$ to time t' or R and $\mathbf{v}$ to time t. It is convenient to choose the first course and write

$$dq = \frac{q}{2\pi}\,d\phi_t = \frac{qK}{2\pi}\,d\phi. \tag{4-46}$$

The combination of Eqs. (4-2) and (4-46) shows the potentials to be

$$\begin{aligned}\Phi(t, \mathbf{r}) &= \frac{q}{4\pi\varepsilon}\oint\frac{d\phi}{R},\\ \mathbf{A}(t, \mathbf{r}) &= \frac{\mu q}{4\pi}\oint\frac{d\phi}{R}\,\mathbf{v},\end{aligned} \tag{4-47}$$

where

$$R^2 = r^2 - 2ar\cos\phi\sin\theta_0 + a^2,$$

$$\mathbf{v} = v(-\hat{x}\sin\phi + \hat{y}\cos\phi),$$

and

$$\mathbf{A} = \frac{\mu q v}{8\pi^2 r}\oint\frac{\cos\phi\,d\phi}{[1-(2a/r)\cos\phi\sin\theta_0+(a^2/r^2)^{1/2}]}.$$

The only term in the vector potential proportional to r^0 through r^{-3} power is

$$\mathbf{A} = \frac{\hat{\phi} q v a\sin\theta_0}{8\pi r^2}. \tag{4-48}$$

The resultant magnetic field is found by taking the curl of

Eq. (4-48). The magnetic moment of the loop is, by definition,

$$\mathbf{m} = \frac{qva}{2}\,\hat{z}\,, \tag{4-49}$$

and combining Eq. (4-49) with the curl of Eq. (4-48) shows the magnetic field to be

$$\mathbf{B} = \frac{\mu m}{4\pi r^3}\,(2\hat{r}\cos\theta_0 + \hat{\theta}\sin\theta_0). \tag{4-50}$$

A comparison of Eqs. (4-45) and (4-50) shows that they are identical in form. It is not possible, therefore, to distinguish between a moment due to separated charges from that due to a current loop by means of the dipolar fields generated.

The force on an electric dipole immersed in an external field $\mathbf{E}_0$ follows from Eq. (2-14) and is

$$\mathbf{F} = (\mathbf{p}\cdot\nabla)\mathbf{E}_0\,. \tag{4-51}$$

Since forces can be calculated by using the Maxwell stress tensor, Eq. (2-18), and since the magnetic and electric fields play dual roles both in Eqs. (2-18) and in (4-45) and (4-50), it follows that the force on a magnetic dipole immersed in an external field $\mathbf{B}_0$ is

$$\mathbf{F} = (\mathbf{p}\cdot\nabla)\mathbf{B}_0\,. \tag{4-52}$$

It is left as an exercise to show that the corresponding resultant torques $\mathscr{T}$ are

$$\begin{aligned} \mathscr{T}_e &= \mathbf{p}\times\mathbf{E}_0\,,\\ \mathscr{T}_m &= \mathbf{m}\times\mathbf{B}_0\,. \end{aligned} \tag{4-53}$$

Exercises

1. Two charges of equal magnitude q are located a distance d apart. Choose a closed surface containing one of the charges and find the force on it by using the Maxwell stress tensor.

2. Verify Eq. (4-51) by integrating the moment of the Maxwell stress tensor over a sphere centered on the dipole.

3. Verify Eq. (4-53) by integrating the moment of the Maxwell stress tensor over a sphere centered on the dipole.

4. A conducting spherical shell of radius a is placed in a uniform electric field $\mathbf{E}$. Find the force tending to separate two halves of the sphere across a diameter plane perpendicular to the electric field.

Ans.: $F = \frac{9}{4}\pi\varepsilon a^2 E^2$.

REFERENCES

Aharoni, J., "The Special Theory of Relativity," 2nd ed., pp. 178–200, 160–165. Oxford Univ. Press (Clarendon), London and New York, 1965.

Jackson, J. D., "Classical Electrodynamics," pp. 464–467. Wiley, New York, 1962.

Kaplan, W., "Advanced Calculus," pp. 199–204. Addison–Wesley, Reading, Massachusetts, 1952.

Landau, L., and E. Lifshitz, "The Classical Theory of Fields," pp. 213–218. Addison–Wesley, Reading, Massachusetts, 1951.

Panofsky, W. K. H., and Phillips, M., "Classical Electricity and Magnetism," 2nd ed., pp. 354–357. Addison–Wesley, Reading, Massachusetts, 1962.

Rohrlich, F., "Classical Charged Particles," pp. 129–134. Addison–Wesley, Reading, Massachusette, 1965.

Schott, G. A., "Electromagnetic Radiation and the Mechanical Reaction Arising from It," pp. 165–172. Cambridge Univ. Press, London and New York, 1912.

5
MULTIPOLAR FIELDS

The Maxwell equations, Eqs. (2-13), give the functional dependence of the electric and magnetic fields on each other. In the absence of any charges, i.e., where ρ_0 and $\mathbf{J}$ are both zero, the equations can be rewritten to show the functional relationship of one vector with itself. This can be done by taking the curl of each of the curl equations and then combining the equations to obtain the result:

$$\nabla \times (\nabla \times \mathbf{E}) + (c^2)^{-1} \frac{\partial^2 \mathbf{E}}{\partial t^2} = 0,$$
$$\nabla \times (\nabla \times \mathbf{B}) + (c^2)^{-1} \frac{\partial^2 B}{\partial t^2} = 0. \tag{5-1}$$

We now make the definition of the vector Laplacian, that

$$\nabla^2 \mathbf{F} = -\nabla \times (\nabla \times \mathbf{F}) + \nabla(\nabla \cdot \mathbf{F}), \tag{5-2}$$

and, using Maxwell's equations, note that Eqs. (5-1) simplify to

$$\nabla^2 \mathbf{E} - (c^2)^{-1} \frac{\partial^2 \mathbf{E}}{\partial t^2} = 0,$$
$$\nabla^2 \mathbf{B} - (c^2)^{-1} \frac{\partial^2 \mathbf{B}}{\partial t^2} = 0. \tag{5-3}$$

Equations (5-3) are the vector wave equations. But the two sets of equations shown there are, of course, not independent, and we wish to obtain a set of independent ones from which to calculate, most easily, the electric and magnetic fields. To this end, we introduce the scalar function $U(t, \mathbf{r})$. $U(t, \mathbf{r})$ is defined to satisfy the scalar wave equation:

$$\nabla^2 U - (c^2)^{-1} \frac{\partial^2 U}{\partial t^2} = 0, \tag{5-4}$$

and it then follows that the vector $\mathbf{r} \times \nabla U$ is a solution of the vector wave equation. The proof goes as follows. The vector Laplacian is calculated with the aid of Eq. (5-2), where

$$\nabla[\nabla \cdot (\mathbf{r} \times \nabla U)] = \nabla[\nabla U \cdot (\nabla \times \mathbf{r}) - \mathbf{r}(\nabla \times \nabla U)],$$

where each term on the right is zero. The remaining term in the Laplacian is

$$\begin{aligned}\nabla \times [\nabla \times (\mathbf{r} \times \nabla U)] &= \nabla \times [\mathbf{r}\nabla^2 U - 2\nabla U - (\mathbf{r} \cdot \nabla)\, \nabla U] \\ &= \mathbf{r} \times \nabla(\nabla^2 U) - \nabla \times [(\mathbf{r} \cdot \nabla)\, \nabla U].\end{aligned}$$

It is a simple exercise to show that

$$(\mathbf{r} \cdot \nabla)\, \nabla U = \nabla[\nabla \cdot (\mathbf{r}U) - 4U].$$

Combining the foregoing gives

$$\begin{aligned}&\nabla^2(\mathbf{r} \times \nabla U) - (c^2)^{-1} \frac{\partial^2}{\partial t^2}(\mathbf{r} \times \nabla U) \\ &\qquad = \mathbf{r} \times \nabla\left[\nabla^2 U - (c^2)^{-1} \frac{\partial^2 U}{\partial t^2}\right] = 0,\end{aligned} \tag{5-5}$$

which completes the proof.

It follows from a comparison of Eqs. (5-3) and (5-5) that the electric field can be calculated by putting it equal to $\mathbf{r} \times \nabla U$, and once $\mathbf{E}$ is known $\mathbf{H}$ can be calculated from it by using Eqs. (2-13). The converse is also true. There are, therefore, the two independent solutions:

$$\begin{aligned}\mathbf{E}^{\mathrm{M}} &= \eta \mathbf{r} \times \nabla U^{\mathrm{M}}, \\ \mathbf{H}^{\mathrm{E}} &= \mathbf{r} \times \nabla U^{\mathrm{E}}.\end{aligned} \tag{5-6}$$

The radial component of the electric field is zero for solutions of the first kind, and these are called TE waves. The radial component of the magnetic field is zero for solutions of the second kind, and these are called TM waves.

To evaluate the fields, we seek to find solutions of Eq. (5-4). We first transform the time dependence into the frequency domain, and then test for solutions. Using the time dependence $e^{i\omega t}$ and using k as defined in Chapter 3, the differential equation to be solved is

$$\nabla^2 U + k^2 U = 0. \tag{5-7}$$

The Laplacian operator ∇^2 involves three independent coordinates. When spherical coordinates are used, the independent coordinates are (r, θ, ϕ). Since each of the coordinates appears in Eq. (5-7), that equation can be solved by splitting it into three separate but complete differential equations and then solving each of the parts by means of a series expansion.

The separation procedure makes use of a theorem applicable in several of the useful coordinate systems, including the Cartesian, cylindrical, and spherical ones. The theorem states that any function can be found by breaking the function into product functions of the three variables taken singly, and then summing over all possible product functions; i.e., one can first express any function $U(r, \theta, \phi)$ as

$$U(r, \theta, \phi) = R(r)Y(\theta, \phi), \tag{5-8}$$

then solve for $R(r)$ and $Y(\theta, \phi)$, and finally form the total function by linear combination of the product functions. Substituting Eq. (5-8) into Eq. (5-7), and multiplying, gives

$$R^{-1}\frac{d}{dr}\left(r^2\frac{dR}{dr}\right) + k^2r^2 = -(Y\sin\theta)^{-1}\frac{\partial}{\partial\theta}\left(\sin\theta\frac{\partial Y}{\partial\theta}\right) - (Y\sin^2\theta)^{-1}\frac{\partial^2 Y}{\partial\phi^2}.$$

All radial dependence appears on the left side of the equation, and all angular dependence on the right side. If values are chosen for the variables θ and ϕ, the right side will have a value independent of r. Repeating the process for r fixed and θ or ϕ variable shows that each side of the equation must be a constant. It is convenient to put that constant equal to $n(n+1)$, where n, itself not yet determined, is the separation constant. The separated equations thus formed are

$$(r^2)^{-1}\frac{d}{dr}\left(r^2\frac{dR}{dr}\right)+\left[k^2-\frac{n(n+1)}{r^2}\right]R=0, \tag{5-9}$$

$$(\sin\theta)^{-1}\frac{\partial}{\partial\theta}\left(\sin\theta\frac{\partial Y}{\partial\theta}\right)+(\sin^2\theta)^{-1}\frac{\partial^2 Y}{\partial\phi^2}+n(n+1)Y=0. \tag{5-10}$$

We first examine the functions $Y(\theta, \phi)$, i.e., the spherical harmonics.

Equation (5-10) can be further subdivided into product functions of the type $Y(\theta, \phi) = \Theta(\theta)F(\phi)$, and separated into two equations with a separation constant m^2:

$$\frac{d^2F}{d\phi^2}+m^2F=0, \tag{5-11}$$

$$(\sin\theta)^{-1}\frac{d}{d\theta}\left(\sin\theta\frac{d\Theta}{d\theta}\right)+\left[n(n+1)-\frac{m^2}{\sin^2\theta}\right]\Theta=0. \tag{5-12}$$

Direct integration of Eq. (5-11) gives

$$F_m(\phi) = A_m \exp(\pm im\phi),$$

and, since $F(\phi)$ must be a single-valued function of position, it must have the same values for ϕ and for $\phi + 2\pi$. Therefore, m must be an integer; and the function can be written as

$$F_m(\phi) = A_m e^{-im\varphi}, \tag{5-13}$$

where m is an integer, either positive or negative, i.e.,

$$m = 0, \pm 1, \pm 2, \pm 3, \cdots. \tag{5-14}$$

To solve Eq. (5-12), it is convenient first to shift the variable to $q = \cos\theta$, thus:

$$\frac{d}{dq}\left[(1-q^2)\frac{d\Theta}{dq}\right]+\left[n(n+1)-\frac{m^2}{1-q^2}\right]\Theta=0. \tag{5-15}$$

Equation (5-15) is a second-order equation, and so will have two solutions. Since one of those solutions has a singularity at $\theta = 0$, it cannot have physical significance in a problem in which all values of θ are present. We are interested only in the non-singular solution, which can be found by a series expansion of Eq. (5-15). The expansion can be simplified by noting that, if $P_n(q)$ is a solution of Eq. (5-15) for $m = 0$, then

$$P_n^{\,m}(q) = (1-q^2)^{m/2}\frac{d^m P_n(q)}{dq^m} \tag{5-16}$$

will be a solution for arbitrary values of m. Hence it is necessary only to solve Eq. (5-15) with $m = 0$, since all solutions can be obtained from $P_n(q)$ by using Eq. (5-16).

We now solve Eq. (5-15) for this case by using the infinite series expansion

$$P_n(q) = \sum_{i=0}^{\infty} a_i q^i$$

and substituting back into the differential equation to obtain

$$0 = \sum_{i=0}^{\infty}[a_{i+2}(i+2)(i+1) - a_i i(i+1) + a_i n(n+1)]q^i,$$

an equation correct for all values of q. This is the case only if the coefficient of each power of q is zero, that is, if

$$\frac{a_{i+2}}{a_i} = \frac{i(i+1)-n(n+1)}{(i+1)(i+2)}. \tag{5-17}$$

A criterion for convergence of an infinite series is that the ratio expressed by Eq. (5-17) approach a number less than unity as i increases without limit. For this case, however, the ratio is unity, and $P_n(q)$ does not converge; hence the expression cannot correspond to a physically realizable field. This conclusion can be evaded, and the ratio of Eq. (5-17) retained, only if the numerator vanishes for some value of i; therefore, all coefficients with subscripts greater than i are zero. This will occur if n is an integer. The function $P_n(q)$ will be even for even n, and odd for odd n, and will be a polynomial of order n.

Values of several functions $P_n{}^m(q)$ are listed in Table 5-1, where $p^2 = 1 - q^2$. These functions are the associated Legendre polynomials. Since these polynomials are of order n, and, from

TABLE 5-1

ASSOCIATED LEGENDRE POLYNOMIALS

n	$m = 0$	$m = 1$	$m = 2$	$m = 3$
0	1	0	0	0
1	q	p	0	0
2	$\frac{1}{2}(3q^2 - 1)$	$3qp$	$3p^2$	0
3	$\frac{1}{2}(5q^3 - 3q)$	$\frac{p}{2}(15q^2 - 3)$	$15qp^2$	p^3
		$p^2 = 1 - q^2$		

Eq. (5-16), the associated polynomials are found by differentiating m times, it follows that there are no solutions for the case $|m| > n$. Therefore, m and n are both integers and subject to the condition $|m| \leq n$.

Equation (5-9) is most easily solved by first extracting from the equation its asymptotic form for large values of r; i.e., as r increases without limit, the equation becomes

$$R'' + k^2 R = 0. \tag{5-18}$$

The solutions of Eq. (5-18) are

$$R(r) = \exp(\pm\, ikr).$$

We therefore put, for any value of radius,

$$R(r) = G_{\pm}(x) e^{\pm ix}, \tag{5-19}$$

where $x = kr$, and then we solve for $G_{\pm}(x)$. The combination of Eqs. (5-9) and (5-19) gives

$$G_{\pm}'' + 2G_{\pm}'(x^{-1} \pm 1) + \left(\frac{\pm 2i}{x} - \frac{n(n+1)}{x^2}\right) G_{\pm} = 0. \tag{5-20}$$

Equation (5-20) can be solved by means of the power series expansion

$$G_{\pm}(x) = \sum_{s=0}^{\infty} c_s x^{s+t}. \tag{5-21}$$

The substitution of Eq. (5-21) into Eq. (5-20) gives, after some algebra, and for n different from zero,

$$t(t+1) = n(n+1), \tag{5-22}$$

$$\frac{c_{s+1}}{c_s} = \frac{\pm 2i(s+t+1)}{n(n+1) - (s+t+1)(s+t+2)}. \tag{5-23}$$

From Eq. (5-22), we see that either $t = n$ or $t = -(n+1)$, and, for these cases, respectively,

$$\frac{c_{s+1}}{c_s} = \frac{\pm 2i(n+s+1)}{n(n+1) - (s+n+1)(s+n+2)}, \tag{5-24}$$

and

$$\frac{c_{s+1}}{c_s} = \frac{\pm 2i(n-s)}{(n-s)(n-s-1) - n(n+1)}. \tag{5-25}$$

The ratio shown in Eq. (5-24), for very large values of s, is $\pm 2i/s$, and the equivalent term in the expansion for $e^{\pm ix}$ is $\pm i/s$. Therefore, the series formed by Eq. (5-24) is not a slowly varying function of radius for large radius, and therefore it is not a satisfactory solution.

Equation (5-25) is the recursion relation for a polynomial of highest order n. The ratios of some selected coefficients are

$$\frac{c_1}{c_0} = \mp i, \qquad \frac{c_2}{c_0} = \frac{(1-n)}{(2n-1)}, \qquad \frac{c_3}{c_0} = \pm \frac{i\,(n-2)}{3(2n-1)}.$$

Some resulting values of $G_{\pm}(x)e^{\pm ix}$ are given in Table 5-2. These functions are the spherical Hankel functions, $h_n^{1,2}(x)$. In what follows, Hankel functions of the first kind are described by the upper sign and Hankel functions of the second kind are described by the lower sign. These functions satisfy the recursion relation

$$h_{n+1} = \frac{2n+1}{x} h_n - h_{n-1} \,. \tag{5-26}$$

In the limit of large radius, the functions go to

$$\lim_{x \to \infty} h_n^{1,2} = \left(\frac{\pm 1}{i} \right)^{n+1} \frac{e^{\pm ix}}{x} \,. \tag{5-27}$$

Therefore, using $e^{i\omega t}$ time dependence, the upper sign represents an incoming wave and the lower sign represents an outgoing wave. In the limit of small radius, the functions go to

$$\lim_{x \to 0} h_n^{1,2} = \mp \frac{i(2n-1)!!}{x^{n+1}} \,, \tag{5-28}$$

where the double factorial indicates

$$(2n-1)!! = (2n-1)(2n-3)(2n-5)\ldots(5)(3)(1). \tag{5-29}$$

The product functions $U(r, \theta, \phi)$, as defined using Eq. (5-8), are, then, products of associated Legendre polynomials and Hankel functions:

$$U^{\mathrm{M}} = \sum_{n=1}^{\infty} \sum_{m=-n}^{m=n} i b_{nm} h_n^{1,2}(x) Y_n{}^m(\theta, \phi), \tag{5-30}$$

$$U^{\mathrm{E}} = \sum_{n=1}^{\infty} \sum_{m=-n}^{m=n} a_{nm} h_n^{1,2}(x) Y_n{}^m(\theta, \phi), \tag{5-31}$$

where

$$Y_n{}^m(\theta, \phi) = \left[\frac{2n+1}{4\pi} \frac{(n-|m|)!}{(n+|m|)!} \right]^{1/2} P_n{}^m(q) e^{-im\phi}, \tag{5-32}$$

where the constants are chosen for later convenience.

The electric and magnetic fields can now be expressed by combining Eqs. (2-13), (5-6), (5-30), and (5-31):

$$
\begin{aligned}
E_r &= i\eta n(n+1)a_{nm}\frac{h_n(x)}{x}Y_n^{\,m}(\theta,\phi),\\
H_r &= n(n+1)b_{nm}\frac{h_n(x)}{x}Y_n^{\,m}(\theta,\phi),\\
E_\theta &= i\eta(ib_{nm}h_nS_n^{\,m} + a_{nm}H_nQ_n^{\,m}), && \text{(5-33)}\\
H_\phi &= -\,ib_{nm}H_nS_n^{\,m} + a_{nm}h_nQ_n^{\,m},\\
E_\theta &= i\eta(b_{nm}h_nQ_n^{\,m} - ia_{nm}H_nS_n^{\,m}),\\
H_\phi &= b_{nm}H_nQ_n^{\,m} + ia_{nm}h_nS_n^{\,m},
\end{aligned}
$$

where

$$H_n(x) = x^{-1}\frac{d}{dx}[xh_n(x)],$$

$$S_n^m = \frac{mY_n^{\,m}}{\sin\theta},$$

$$Q_n^m = \frac{dY_n^{\,m}}{d\theta}.$$

TABLE 5-2
SPHERICAL HANKEL FUNCTIONS

n	h_n
0	$\frac{\pm 1}{ix}\exp(\pm ix)$
1	$x^{-1}\left(\frac{\pm 1}{ix} - 1\right)\exp(\pm ix)$
2	$\frac{\pm i}{x}\left(1 \pm \frac{3i}{x} - \frac{3}{x^2}\right)\exp(\pm ix)$
3	$x^{-1}\left(1 \pm \frac{6i}{x} - \frac{15}{x^2} \mp \frac{i15}{x^3}\right)\exp(\pm ix)$

REFERENCE

Stratton, J. A., "Electromagnetic Theory," Chap. VII. McGraw-Hill, New York, 1941.

6

ENERGY STORAGE AND FLOW IN MULTIPOLAR FIELDS

The flow of energy, momentum, and angular momentum from a multipolar source can be found by using the field equations, i.e., from Eqs. (4-29) and (4-33). However, the development of the force equations, Eqs. (4-21) and (4-22), depends upon Eqs. (4-9) and (4-18). However, it is apparent that they are valid only in a nonaccelerating frame. We, therefore, now seek a method of accounting for the reaction force due to the acceleration term and examine results of the additional force.

We begin the discussion with the power flow from a volume that contains only electric multipoles. For this case, the radial portion of the average value of N is, using Eqs. (5-33),

$$N_r = \frac{i\eta}{2} \sum_{n=1}^{\infty} \sum_{n'=1}^{\infty} \sum_{m=-n}^{n} \sum_{m'=-n'}^{n'} a_{nm} a^*_{n'm'} \cdot H_n(x) h^*_{n'}(x) (Q_n{}^m Q_{n'}^{m'*} + S_{n'}^{m'} S_{n'}^{m'}). \tag{6-1}$$

Now the average power radiated from the source is

$$P = \oint r^2 N_r \, d\Omega. \tag{6-2}$$

When Eq. (6-1) is substituted into Eq. (6-2), and the result is combined with Eq. (7A-12) from the Appendix,

$$\oint d\Omega (Q_n{}^m Q_{n'}^{m'*} + S_n{}^m S_{n'}^{m'*}) = n(n+1)\, \delta_{m,m'}\, \delta_{n,n'}, \tag{6-3}$$

the power radiated is found to be

$$P^{\mathrm{E}} = \frac{i\eta}{2k^2} \sum_{n=1}^{\infty} \sum_{m=-n}^{m=n} n(n+1) a_{nm} a_{nm}^* x^2 H_n(x) h_n^*(x). \qquad (6\text{-}4)$$

A similar analysis when only magnetic multipoles exist shows that the radiated power is then

$$P^{\mathrm{M}} = \frac{\eta}{2ik^2} \sum_{n=1}^{\infty} \sum_{m=-n}^{n} n(n+1) b_{nm} b_{nm}^* H_n^*(x) h_n(x) x^2. \qquad (6\text{-}5)$$

From Eqs. (6-4) and (6-5), we see that power is absorbed or emitted in proportion to the sum of the squares of the coefficients a_{nm} and b_{nm}. Since emitted power is positive, Hankel functions of the first kind describe power absorption, and Hankel functions of the second kind describe power emission.

Next we seek to find the average reactive energy that is stored in the radiation field external to the radiating volume, and we first find that energy for the electric multipoles. For this purpose, we introduce a generalized forcing function and a generalized flow:

$$P = \frac{1}{2} \sum_{n=1}^{\infty} V_n I_n^*, \qquad (6\text{-}6)$$

and, after comparing Eqs. (6-4) and (6-6), we put

$$V_n(x) = \left(\frac{\eta n(n+1)}{k^2}\right)^{1/2} i a_{nm} x H_n(x),$$

$$I_n(x) = \left(\frac{\eta n(n+1)}{k^2}\right)^{1/2} a_{nm} x h_n(x). \qquad (6\text{-}7)$$

The ratio of Eqs. (6-7) gives a generalized impedance for electric multipoles:

$$Z_n^{\mathrm{E}} = \frac{V_n(x)}{I_n(x)} = \frac{iH_n(x)}{h_n(x)}. \qquad (6\text{-}8)$$

A similar analysis for magnetic multipolar sources shows their equivalent impedance to be

$$Z_n^{\mathrm{M}} = \frac{h_n(x)}{iH_n(x)}. \qquad (6\text{-}9)$$

The impedances are ratios of polynomials of highest order

$n + 1$. They can, therefore, be exactly synthesized using conventional circuit synthesis techniques. We consider Z_1^{E} as an example. From Table 5-2,

$$h_1 = -x^{-1}[1 + (ix)^{-1}]e^{-ix}$$

for outgoing waves. It follows from the definition of $H_n(x)$ that

$$H_1 = \frac{i}{x}\left[1 - \frac{i}{x} - (x^2)^{-1}\right]e^{-ix},$$

and so the impedance, using Eq. (6-8), is

$$_eZ_1 = (ix)^{-1} + [1 + (ix)^{-1}]^{-1}, \tag{6-10}$$

and similarly, using Eq. (6-9),

$$_mZ_1 = \{(ix)^{-1} + [1 + (ix)^{-1}]^{-1}\}^{-1}. \tag{6-11}$$

The equivalent circuits for Eqs. (6-10) and (6-11) and for an arbitrary n electric multipole are shown in Fig. 6-1. The resistances are positive for energy absorption. A similar analysis for emission reveals similar circuits save the resistances are negative.

The relationship between V and I for the electric dipolar case ($n = 1$) follows from the circuit. Let the current through the inductor be I_1 and the current through the resistor be I_2; then, in terms of the circuit parameters R, L, and C,

$$I_2 = \frac{L}{R}\frac{dI_1}{dt} \tag{6-12}$$

and

$$I_1 = I - I_2. \tag{6-13}$$

After repeated substitutions of Eqs. (6-12) and (6-13) into each other, the result is

$$I_1 = \sum_{i=0}^{\infty}\left(-\frac{L}{R}\right)^i\frac{d^iI}{dt^i}. \tag{6-14}$$

Also, from the same circuit,

$$V = C^{-1}\int I\,dt + L\frac{dI_1}{dt},$$

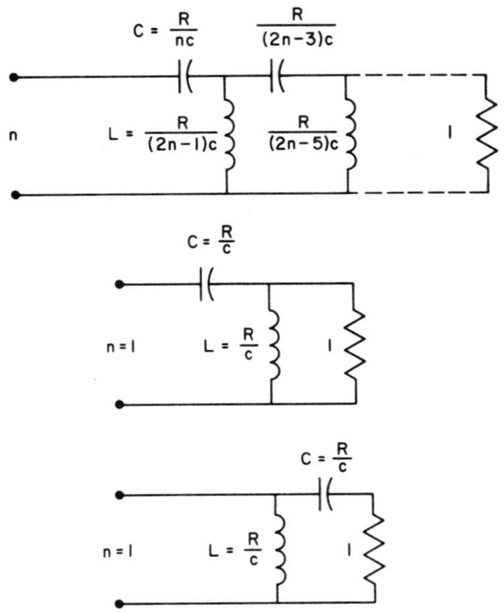

FIG. 6-1. Equivalent circuit of multipole radiators.

so that the voltage-current relationship is

$$V = C^{-1} \int I\,dt + L \sum_{i=0}^{\infty} \left(- \frac{L}{R} \right)^{i} \frac{d^{i+1}I}{dt^{i+1}}. \tag{6-15}$$

A similar analysis for the $n = 2$ case reveals that the differential equation corresponding to Eq. (6-15) is

$$V = \frac{2}{C} \int I\,dt + \frac{L}{3} \frac{dI}{dt} - \frac{CL^2}{9} \frac{d^3I}{dt^3} + \frac{RC^2L^2}{9} \frac{d^4I}{dt^4} + \frac{C^2L^3}{27} \left(1 - \frac{3CR^2}{L} \right) \frac{d^5I}{dt^5} + \cdots. \tag{6-16}$$

Next we assume that the radiated power arises from the motion of a point charge q as it moves within the source. For simplicity, we consider it to be confined to movements along the y axis. The power supplied to the charge from an external field is, then,

$$P = \frac{qE}{2}\frac{dy^*}{dt}, \tag{6-17}$$

where y is the position of the charge. The mechanical reaction to this power is the sum of inertial effect, electrostatic field effect, and radiated power effect as calculated from Eq. (6-4). A direct comparison between Eqs. (6-6) and (6-17) can be made by substituting

$$qE = uV,$$

$$\frac{dy}{dt} = \frac{1}{u} I,$$

and noting that, therefore, the differential equation, Eq. (6-15), becomes, in terms of E and y and the circuit parameters of Fig. 6-1,

$$qE = u^2\left[\frac{cy}{R} + \frac{R}{c}\sum_{n=0}^{\infty}\left(-\frac{R}{c}\right)^n \frac{d^{n+2}y}{dt^{n+2}}\right]. \tag{6-18}$$

To evaluate u, we note from Eq. (6-18) that the energy for $\omega = 0$ is

$$W = \frac{cy^2u^2}{2R},$$

and from Eqs. (6-15) and (6-7), and from Eqs. (4-45), (5-32), and (5-33),

$$W = \frac{cII^*}{2\omega^2 R} = \frac{q^2y^2}{12\pi\varepsilon R^3},$$

to determine u^2 as

$$u^2 = \frac{\eta q^2}{6\pi R^2}, \tag{6-19}$$

and thereby to determine the total differential equation

$$qE = \frac{q^2y}{6\pi\varepsilon R^3} + \frac{q^2\mu}{6\pi R}\sum_{n=0}^{\infty}\left[\left(-\frac{R}{c}\right)^n \frac{d^{n+2}y}{dt^{n+2}}\right]. \tag{6-20}$$

The ratio of the coefficients for successive terms under the summation is the ratio between the radius of the region containing the moving charge and the wavelength of the radiation.

By a similar process, the differential equation for $n = 2$ is found to be

$$qE = \frac{q^2}{3\pi R}\left\{\left(\frac{c}{R}\right)^2 \frac{y}{6} + \frac{1}{6}\frac{d^2y}{dt^2} - \left(\frac{1}{18}\right)\left(\frac{R}{6}\right)^2 \frac{d^4y}{dt^4} + \frac{1}{18}\left(\frac{R}{c}\right)^3 \frac{d^5y}{dt^5} - \frac{1}{27}\left(\frac{R}{6}\right)^4 \frac{d^6y}{dt^6} + \cdots\right\}. \tag{6-21}$$

The lowest-order term describing power loss is in proportion to the fifth-order derivative with respect to time, instead of the third-order effect resulting from dipolar radiation.

Equation (6-20), of course, describes only the radiation reaction, i.e., the force due to the second term on the right of Eq. (4-6). The first term on the right is accounted for by Eq. (4-21) and the configuration force. For speed much less than c, and for potential energy $I(r)$, these added terms are

$$m\frac{d^2y}{dt^2} - \nabla I(r).$$

When these terms are added, and the possibility of dipolar moments in each of the three orthogonal directions is incorporated, the total differential equation of motion becomes

$$q\mathbf{E} = \frac{q^2\mathbf{r}}{6\pi\varepsilon R^3} - \nabla I(\mathbf{r}) + m\frac{d^2\mathbf{r}}{dt^2} + \frac{\mu q^2}{6\pi R}\sum_{n=0}^{\infty}\left(-\frac{R}{c}\right)^n \frac{d^{n+2}\mathbf{r}}{dt^{n+2}}. \tag{6-22}$$

Equation (6-22) is the complete differential equation for the dynamic response of a point charge that is radiating in a dipolar mode, to a driving field.

To analyze the results of this equation, we confine our attention to electrons so that $q = -e$, and we also confine our attention to cases that satisfy the inequalities

$$kR \ll 1,$$
$$\frac{a}{R} \ll 1, \tag{6-23}$$

and apply Eq. (6-22) to a one-dimensional harmonic oscillator. The result is the truncated equation

$$-eE = m(\omega_1{}^2 + \omega_r{}^2)y + m\frac{d^2y}{dt^2} - \frac{2ma}{3c}\frac{d^3y}{dt^3}, \tag{6-24}$$

where $\omega_r{}^2 = 2ac^2/(3R^3)$. The difference between Eq. (6-24) and a similar one that does not account for the reaction field is, of course, the term $m\omega_r{}^2y$. The resonant frequency of the radiator is

$$\omega_0 = (\omega_1{}^2 + \omega_r{}^2)^{1/2}, \tag{6-25}$$

and so it satisfies the inequality

$$\omega_0 > \left(\frac{2ac^2}{3R^3}\right)^{1/2} \tag{6-26}$$

independently of the potential. Therefore, an undriven oscillator can only radiate at a wavelength smaller than that given in inequality (6-27):

$$\lambda < \left(\frac{6\pi^2R^3}{a}\right)^{1/2} = 1.45 \times 10^{-7}R_0^{3/2}, \tag{6-27}$$

where R_0 is the radius of the spherical shell measured in Angstrom units, 10^{-10}m.

The maximum displacement of the charge, y_0, is equal to the radius R of the region. Therefore, for dipolar radiation to occur, and using Eq. (6-24), the energy

$$W_1 = \frac{q^2}{12\pi\varepsilon R} \tag{6-28}$$

must be stored in the reactive field. So energy in excess of that shown in Eq. (6-28) must be available for energy to be radiated into the far field. Equations (6-4) and (6-5) show comparative electric and magnetic power flows. They are complex conjugate functions if $|a_{nm}| = |b_{nm}|$. And with conjugate power flows, the respective reactive energy is out of phase. Therefore, if energy W_1, as given by Eq. (6-28), is being exchanged between the field and the source by electric dipolar radiation, then magnetic dipolar radiation with $|b_{nm}| = |a_{nm}|$ can simultaneously exist without additional reactive energy. Indeed, unless the coefficients are equal in magnitude, the source must store the reactive energy during a portion of each cycle.

Returning to Eq. (6-24) in its homogeneous form,

$$-\frac{2a}{3c}\frac{d^3y}{dt^3} + \frac{d^2y}{dt^2} + \omega_0{}^2 y = 0, \tag{6-29}$$

and, since $ka \ll 1$, it follows that y is given, very nearly, by

$$y = y_0 \exp\left[\left(i\omega_0 - \frac{\gamma}{2}\right)t\right]$$

where

$$\gamma = \frac{2a}{3c}\,\omega_0{}^2.$$

From Eq. (6-29), the average potential energy of the charge is

$$W_{\mathrm{p}} = \frac{m}{4}\,\omega_0{}^2 y_0{}^2\, e^{-\gamma t}, \tag{6-30}$$

and so the power loss by radiation is

$$P = \frac{2a}{3c}\,\omega_0{}^2 W_{\mathrm{p}}. \tag{6-31}$$

The radiative Q, which, by definition, is

$$Q = \frac{\omega W_{\mathrm{p}}}{P},$$

is given by

$$Q = \frac{3c}{2a\omega}. \tag{6-32}$$

Similarly, in resonant devices, the relationship between Q and bandwidth $\Delta\lambda$ is that

$$Q = \frac{\lambda}{\Delta\lambda},$$

and so the bandwidth is

$$\Delta\lambda = \tfrac{4}{3}\pi a = 11.8 \times 10^{-15} \quad \text{m}, \tag{6-33}$$

a result that is independent of the wavelength.

The decay time τ of an oscillator is the time for the stored energy to drop to $1/e$ of its initial value, and is given by Q/ω, or

$$\tau = \frac{3\lambda^2}{8\pi^2 ac} < \frac{9R^3}{4a^2c} = 9.5 \times 10^{-10} R_0{}^3, \tag{6-34}$$

where R_0 is measured in Angstrom units (10^{-10} m), a result with important ramifications for energy absorption.

By use of the reciprocity theorem, these results can be applied to the atom viewed as a receiving atenna. The reciprocity theorem follows from the Maxwell equations. To develop it, let $\mathbf{E}_m$, $\mathbf{H}_m$ and $\mathbf{E}_n$, $\mathbf{H}_n$ be the fields inside any closed region originating in two sources, say m and n. We begin by considering the divergence at each point in space:

$$\nabla \cdot (\mathbf{E}_m \times \mathbf{H}_n) = -i\omega\mu\mathbf{H}_m \cdot \mathbf{H}_n - i\omega\varepsilon\mathbf{E}_m\mathbf{E}_n.$$

The equality follows from the Maxwell equations and the use of exponential time dependence. Upon exchanging subscripts, it follows that

$$\nabla \cdot (\mathbf{E}_m \times \mathbf{H}_n - \mathbf{E}_n \times \mathbf{H}_m) = 0,$$

and integrating over all space shows that

$$\int (\mathbf{E}_m \times \mathbf{H}_n - \mathbf{E}_n \times \mathbf{H}_m) \cdot d\mathbf{S} = 0. \tag{6-35}$$

Equation (6-35) is a form of the reciprocity theorem. To apply it to a specific model, let us assume that m and n represent two different antennas. They can be of arbitrary and different shapes, but for simplicity we assume that each is constructed of perfect conductors and fed by a coaxial line. We wish to take the surface integrals in Eq. (6-35) and evaluate them over the surface enclosing each antenna and its source. Over the perfectly conducting portion of the surface, $\mathbf{E} \times \mathbf{dS} = 0$; so the surface integrals need be evaluated only within the coaxial lines, or

$$\int_1 (\mathbf{E}_m \times \mathbf{H}_n - \mathbf{E}_n \times \mathbf{H}_m) \cdot d\mathbf{S} + \int_2 (\mathbf{E}_m \times \mathbf{H}_n - \mathbf{E}_n \times \mathbf{H}_m) \cdot d\mathbf{S} = 0, \tag{6-36}$$

where 1 and 2 represent the surface areas within the two coaxial lines. Now within the lines, $\mathbf{H} = \hat{\phi} I/(2\pi r)$, $\mathbf{E} = \hat{\mathbf{r}} E_r$, and $V = \int_{R_i}^{R_0} E_r \, dr$. We evaluate the first term in Eq. (6-36) to obtain

$$\int_1 (\mathbf{E}_m \times \mathbf{H}_n) \cdot d\mathbf{S} = 2\pi \int_{R_i}^{R_0} E_{rm} H_{\phi n} r \, dr = I_{1n} \int_{R_i}^{R_0} E_{rm} \, dr = I_{1n} V_{1m}.$$

Evaluating each integral in Eq. (6-36) in the same manner yields

$$V_{1m} I_{1n} + V_{2m} I_{2n} - V_{1n} I_{1m} - V_{2n} I_{2m} = 0. \tag{6-37}$$

We wish to write the currents as functions of the applied potentials as follows:

$$\begin{aligned} I_1 &= Y_{11} V_1 + Y_{12} V_2, \\ I_2 &= Y_{21} V_1 + Y_{22} V_2. \end{aligned} \tag{6-38}$$

Substituting Eqs. (6-38) into Eq. (6-37) yields

$$(V_{1m} V_{2n} - V_{2m} V_{1n})(Y_{12} - Y_{21}) = 0. \tag{6-39}$$

Since the potentials V_m and V_n can be controlled independently of each other, $Y_{12} = Y_{21}$. Therefore, an ideal antenna will have the same radiation pattern and radiation impedance for receiving as for transmitting, and so these values calculated for a transmitting antenna apply to the receiving antenna, and vice versa.

Next we examine the effective cross-sectional area of a receiving antenna. To do so, we consider the antenna as a radiator driven by a voltage source in series with an impedance that is the complex conjugate of the antenna impedance (see Fig. 6-2) and then invoke the reciprocity theorem.

Figure 6-2 shows that the power dissipated in the source resistance is

$$P_{\mathrm{d}} = \frac{V^2}{8R_r},$$

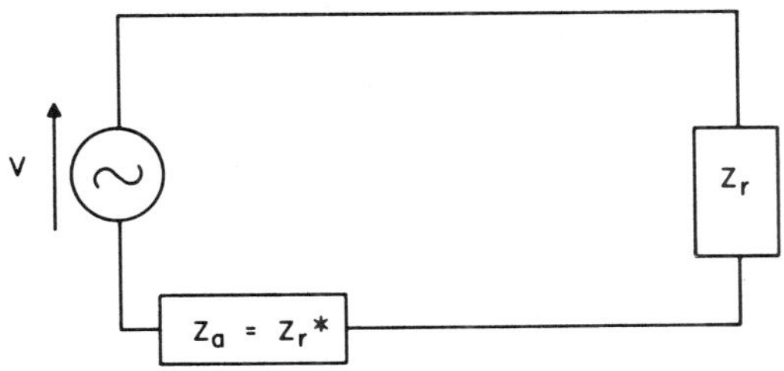

FIG. 6-2. The antenna as a receiving unit attached to a conjugately matched impedance (Z_r = impedance of receiver, Z_a = impedance of antenna).

where V is the voltage of the generator. The effective aperture of an antenna is the ratio between two magnitudes: that of the power dissipated by the load resistance as a result of voltage induced in the antenna by an incident plane wave, and that of the power-flow density in the wave. That is, if V is the voltage induced in the receiving antenna, the aperture A_e is

$$A_e = \frac{VV^*}{8NR_r}, \tag{6-40}$$

where, as usual, N is the Poynting vector in the incident wave. We note that A_e is equivalent to the area of the incident wavefront from which all power is dissipated in the load impedance. For $y_0 \ll \lambda$, the voltage induced in the antenna is $y_0 E$. The Poynting vector is E^2/η. The radiation resistance is determined by the power loss into the radiation field, and by definition it is proportional to the power radiated as

$$P_d = \tfrac{1}{4} R_r I^2 = \tfrac{1}{4} R_r \omega^2 q^2. \tag{6-41}$$

By using Eqs. (6-31) and (6-41), the radiation resistance is found to be

$$R_r = \frac{2\pi\eta}{3}\left(\frac{y_0}{\lambda}\right)^2. \tag{6-42}$$

Upon combining all terms, the effective aperture is found to be

$$A_e = \frac{3\lambda^2}{8\pi}. \tag{6-43}$$

That is, so long as the antenna is resonant, the power is extracted from an area that is in proportion to the square of the wavelength and that depends in no way upon the actual physical size of the antenna. The limiting factor for power absorption is that the required reactive field energy increases with wavelength, and with it the resonant frequency of the system. For this reason, there exists a maximum time for power absorption or emission which is dependent upon the size of the radiator.

Exercise

1. Show that Fig. 6-1 is correct for $n = 2$ by synthesizing the equivalent circuit.

REFERENCES

Chu, L. J., "Physical Limitations of Omnidirectional Antennas," *J. Appl. Phys.* **19,** 1163–1175 (1948).

Jenkins, F. A., and White, J. E., "Fundamentals of Optics," 3rd ed., pp. 443–444. McGraw-Hill, New York, 1957.

Lorentz, H. A., "The Theory of Electrons," 2nd ed., pp. 259–260. Dover, New York, 1952.

Panofsky, W. K. H., and Phillips, M., "Classical Electricity and Magneticism," 2nd ed., pp. 401–403. Addison–Wesley, Reading, Massachusetts, 1962.

7

ENERGY AND MOMENTUM FLOW IN MULTIPOLAR FIELDS

Energy and momentum flows from closed regions can be found by using the Poynting's vector, Eq. (4-29), and the definition of angular momentum. Each on a per unit time basis, the net radiated quantities are as follows. The energy radiated is

$$W = \oint d\mathbf{S} \times \mathbf{N}; \tag{7-1}$$

the radiated momentum is

$$\mathbf{G} = c^{-1} \oint dS\, \mathbf{N}; \tag{7-2}$$

and the radiated angular momentum is

$$\mathbf{L} = c^{-1} \oint dS\, \mathbf{r} \times \mathbf{N}. \tag{7-3}$$

To evaluate the net flows, we are concerned only with the far-field limits of the radiation. For that case, Eqs. (5-33) become

$$E_r = i\eta(\mp i)^{n+1} n(n+1) a_{nm} Y_n{}^m \frac{e^{\pm ix}}{x^2},$$

$$H_r = (\mp i)^{n+1} n(n+1) b_{nm} Y_n{}^m \frac{e^{\pm ix}}{x^2},$$

$$E_\theta = \eta(\mp i)^{n+1}(-b_{nm} S_n{}^m \mp a_{nm} Q_n{}^m) \frac{e^{\pm ix}}{x},$$

$$H_\phi = (\mp i)^{n+1}(\pm b_{nm} S_n{}^m + a_{nm} Q_n{}^m) \frac{e^{\pm ix}}{x},$$

$$E_\phi = i\eta(\mp i)^{n+1}(b_{nm}Q_n{}^m \pm a_{nm}S_n{}^m)\frac{e^{\pm ix}}{x},$$

$$H_\theta = i(\mp i)^{n+1}(\pm\ b_{nm}Q_n{}^m + a_{nm}S_n{}^m)\frac{e^{\pm ix}}{x}. \tag{7-4}$$

In the far fields that carry energy or momentum, the radial and azimuthal components of the Poynting vector are of interest. These follow from Eqs. (7-4) and are

$$\begin{aligned} N_r &= \frac{-\eta}{2x^2}[\pm(S_n{}^m S_n^{m*} + Q_n{}^m Q_n^{m*})(a_{nm}a_{nm}^* + b_{nm}b_{nm}^*) \\ &\quad + (S_n{}^m Q_n^{m*} + S_n^{m*}Q_n{}^m)(a_{nm}^* b_{nm} + a_{nm}b_{nm}^*)], \\ N_\phi &= \frac{\eta n(n+1)}{2x^3}[(a_{nm}^* a_{nm} Y_n{}^m S_n^{m*} + b_{nm}b_{nm}^* Y_n^{m*}S_n{}^m) \\ &\quad \pm a_{nm}b_{nm}^*(Y_n{}^m Q_n^{m*} + Y_n^{m*}Q_n{}^m)]. \end{aligned} \tag{7-5}$$

The combinations of Eqs. (7-1), (7-2), (7-3), and (7-4) shows the far-field limit of the energy and the z components of the momentum to be

$$\begin{aligned} W &= r^2 \int d\Omega\, N_r, \\ G_z &= \frac{r^2}{c}\int d\Omega\, N_r \cos\theta, \\ L_z &= \frac{r^3}{c}\int d\Omega\, N_\phi \sin\theta. \end{aligned} \tag{7-6}$$

The combination of Eqs. (7-5) and (7-6) shows the results, when only one value of n is present in the fields, that

$$\begin{aligned} W &= \mp\frac{\eta}{2k^2}\sum_{m=-n}^{n} n(n+1)(a_{nm}a_{nm}^* + b_{nm}b_{nm}^*), \\ G_z &= \frac{\eta}{2ck^2}\sum_{m=-n}^{n} m(a_{nm}b_{nm}^* + a_{nm}^* b_{nm}), \\ L_z &= \frac{\eta}{2ck^3}\sum_{m=-n}^{n} n(n+1)m(a_{nm}a_{nm}^* + b_{nm}b_{nm}^*), \end{aligned} \tag{7-7}$$

where the integrals of Eqs. 7-6 were evaluated by using Eqs. (7A-3), (7A-5), (7A-7), (7A-8), (7A-9), and (7A-10).

We next consider a plane wave traveling in the $+z$ direction and incident from $-\infty$ that is polarized in the x direction. Each component depends upon spatial position as

$$\exp(-ikr\cos\theta).$$

To compare Eqs. (5-33) with the incoming plane wave better, we next express this exponential in terms of spherical functions. The expansion developed in Chapter 5 for the fields about any origin is a general one that must include all possibilities of fields, and, therefore, it includes a plane wave passing through the region. But for that case, the fields are nonsingular at the origin, and so we must form the spherical Bessel functions $j_n(x)$ by taking

$$j_n(x) = \tfrac{1}{2}[h_n{}^1(x) + h_n{}^2(x)] \tag{7-8}$$

and we put

$$\exp(-ix\cos\theta) = \sum_{n=0}^{\infty} C_n j_n(x) Y_n(\cos\theta), \tag{7-9}$$

where we must evaluate the infinite set of coefficients C_n. To do so, we begin by multiplying Eq. (7-9) by $Y_m(\cos\theta)$ and integrating over all solid angles. The result is

$$C_m j_m(x) = \oint d\Omega\; Y_m \exp(-ikr\cos\theta). \tag{7-10}$$

Next we differentiate both sides of Eq. (7-10) n times and then evaluate the coefficients in the limit as x becomes vanishingly small. In that limit, from Table 5-2 with extensions, the function $j_n(x)$ becomes equal to

$$\lim_{x\to 0} j_n(x) = \frac{x^n}{(2n+1)!!}, \tag{7-11}$$

and the nth differential of Eq. (7-10) becomes

$$\frac{n!C_n}{(2n+1)!!} = (-i)^n \oint q^n Y_n(q)\, d\Omega. \tag{7-12}$$

Since it follows from the theory of Legendre polynomials that

$$\oint q^n Y_n(q)\, d\Omega = \left(\frac{4\pi}{2n+1}\right)^{1/2} \frac{(n!)^2 2^{n+1}}{(2n+1)!},$$

the set of coefficients is given by

$$C_n = (-i)^n 2\left(\frac{4\pi}{2n+1}\right)^{1/2}. \tag{7-13}$$

Now the electric and magnetic field vectors are directed in the x and y directions, respectively, and

$$\begin{aligned} \hat{x} &= \hat{r} \sin\theta \cos\phi + \hat{\theta} \cos\theta \cos\phi - \hat{\phi} \sin\phi, \\ \hat{y} &= \hat{r} \sin\theta \sin\phi + \hat{\theta} \cos\theta \sin\phi + \hat{\phi} \cos\phi, \end{aligned} \tag{7-14}$$

so that the radial components of the fields are

$$\begin{aligned} E_\mathrm{r} &= E_0 \exp(-ix\cos\theta) \sin\theta \cos\phi, \\ H_\mathrm{r} &= \frac{E_0}{\eta} \exp(-ix\cos\theta) \sin\theta \sin\phi, \end{aligned} \tag{7-15}$$

where $x = kr$ as before. Now, since, from Eq. (5-16),

$$\frac{dP_n(\cos\theta)}{d\theta} = -P_n{}^1$$

and

$$\sin\theta \exp(-ix\cos\theta) = (ix)^{-1} \frac{\partial}{\partial\theta} [\exp(-ix\cos\theta)],$$

Eqs. (7-9) and (7-15) can be combined to give

$$\begin{aligned} E_{\mathrm{r}nm} &= iE_0(-i)^n \left(\frac{4\pi(n+1)!}{(2n+1)(n-1)!}\right)^{1/2} \frac{j_n(x)\, Y_n{}^m}{x}, \\ H_{\mathrm{r}nm} &= \frac{mE_{\mathrm{r}nm}}{i\eta}, \qquad m = \pm 1, \end{aligned} \tag{7-16}$$

where we have used Eq. (5-32). The fields are 0 for $m \neq \pm 1$.

Now a comparison of Eqs. (5-31) and (7-16) shows that

$$E_{\mathrm{r}n} = \frac{i\eta n(n+1)}{x} U_n{}^\mathrm{E},$$

$$H_{rn} = -\frac{in(n+1)}{x} U_n^{\mathrm{M}}. \tag{7-17}$$

A comparison of Eqs. (5-30), (5-31), and (7-17) shows that a plane wave is the special case where the a_{nm}, b_{nm} coefficients of Eqs. (5-33) become equal to

$$\begin{aligned} A_{nm} &= \frac{E_0}{\eta}(-i)^n \left[\frac{4\pi(n+1)!}{(2n+1)(n-1)!}\right]^{1/2}, \\ B_{nm} &= -mA_{nm}, \qquad m = \pm 1, \\ A_{nm} &= B_{nm} = 0, \qquad m \neq \pm 1. \end{aligned} \tag{7-18}$$

Comparison of Eqs. (5-33) and (7-11), however, shows that the driving field at the surface of a small sphere varies as x^{n-1}. Since the coefficients a_n that create the radiated field multiply R^n for small values of kR, we retain only the $n = 1$ terms.

Since the field coefficient $A_{10} = 0$, we put the corresponding term equal to zero in the generated field, or

$$a_{10} = 0. \tag{7-19}$$

Since a plane wave transports energy with an energy-to-momentum ratio of c, the ratio of absorbed energy to momentum must be

$$\frac{W}{G_z} \leqslant c. \tag{7-20}$$

The combination of Eqs. (7-7) and (7-20) shows that

$$1 \geqslant \frac{2\sum_{m=-1}^{m=1} \pm (a_{1m}a_{1m}^* + b_{1m}b_{1m}^*)}{\sum_{m=-1}^{m=1} m(a_{1m}b_{1m}^* + a_{1m}^*b_{1m}^*)},$$

a condition which cannot be satisfied unless the double sign in the numerator differs between the $m = +1$ and $m = -1$ cases. That is, emission must occur simultaneously with absorption, and vice versa. But using Eq. (7-7), remembering Eq. (7-19), that choice of sign shows that

$$\frac{W}{L_z} = \pm\,\omega. \tag{7-21}$$

The ratio of exchanged energy to angular momentum is equal to the radian frequency. In many instances, the exchanged energy is in proportion to the frequency, or

$$W = \pm \hbar\omega. \tag{7-22}$$

see Eq. (12-14), for which case the radiation will carry angular momentum

$$L_z = \pm \hbar. \tag{7-23}$$

Therefore, the change in angular momentum of the source will also be $\hbar$.

Now the coefficients b_n in a generated field are of order R^{n+1}, and so b_n is usually ignored and radiation due to only a_n is considered. However, a near-field reactive energy must exist for energy to be radiated away by electric multipolar radiation, as given by Eq. (6-28) for $n = 1$. And if that energy exists, then magnetic dipolar radiation is driven and can occur with coefficients in proportion to those of Eq. (7-18),

$$b_{nm} = - m a_{nm} \tag{7-24}$$

without additional reactive energy. Indeed, Eq. (7-20) can only be satisfied if b_{nm} is different from zero, for a single value of n. The reactive energy of Eq. (6-28) is minimized for a given power flow when Ineq. (7-20) is an equality, as can be easily seen. We therefore take (7-20) to be an equality.

Equations (7-20) and (7-24) are then satisfied, and a net momentum of W/c is extracted from the plane wave, if

$$\begin{aligned} a_{11} &= b_{11}, \\ a_{1\,-1} &= b_{1\,-1}, \\ a_{1\,-1} &= \sqrt{3}\, a_{11} \end{aligned} \tag{7-25}$$

These conditions are also satisfied for an exchange of $m = +1$ with $m = -1$. The radiation pattern of the absorbed and emitted power follow from Eqs. (7-5), (7-19), and (7-25). They are shown in Figure 7-1.

The condition expressed by Eq. (7-20) is an important one,

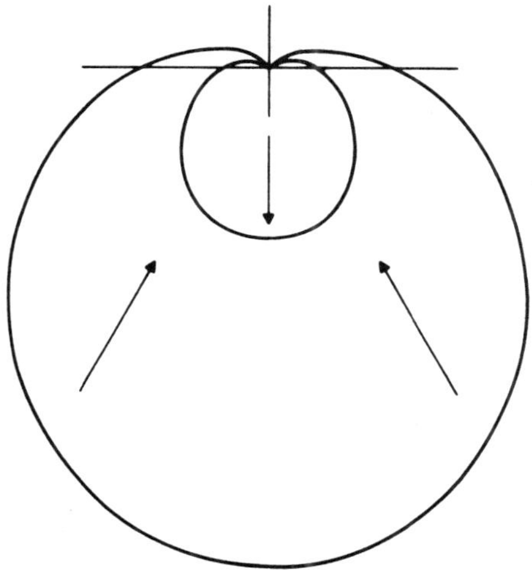

FIG. 7-1. Radiation pattern for classical photon.

since Einstein, in an historic paper, assumed that the equality held for energy absorption from a plane wave, without stipulating how, but then went on to assume that momentum could not be emitted by a classical dipole radiator.[1] He went on to show, from a statistical argument, that the equality must hold for emission, and he was thereby driven to conclude that a quantum theory of radiation was unavoidable.[2] However, we now see that the necessary momentum condition can be met by classical dipolar radiators. Therefore, Einstein's result does not specifically require a quantum theory of radiation.

[1] "Bei der Energieabgabe durch Ausstrahlung wird im Falle des Planckschen Resonators im ganzen kein Impuls auf den Resonator übertragen, weil nach der klassischen Theorie die Ausstrahlung in einer Kugelwelle erfolgt."

[2] "Diese durch die"—momentum considerations—"gerforderten Eigenschaften der Elementarprozesse lassen die Aufstellung einer eigentlich quantenhaften Theorie der Strahlung fast unvermeidlich eschenien."

The increasing use of miniaturized antennas makes Eqs. (7-20) of interest to their further development, for it places an additional constraint upon them. Although the momentum condition is nearly a trivial one for antennas whose dimensions are about the size of a wavelength, the condition cannot be easily met for small antennas. For then the antenna must have sufficient internal degrees of freedom to support a magnetic dipolar field simultaneously with the desired electric dipolar one. (An electric quadropolar field will also satisfy the momentum condition but requires an additional reactive energy.) The required magnitudes of the fields increase inversely with the size of the radiator [see Eq. (6-28)]. Unless this condition is met, the energy extracted from a plane wave will decrease with the physical size of the antenna.

Appendix 7A

The functions $Y_n^m(\theta, \phi)$ are defined by Eq. (5-32). S_n^m and Q_n^m are defined just below Eqs. (5-33). It can be easily verified by substituting back into differential equation (5-12) that P_n^m satisfies the relationship

$$P_n^m(q) = \frac{(1 - q^2)^{m/2}}{2^n n!} \frac{d^{n+m}(q^2 - 1)^n}{dq^{n+m}} . \tag{7A-1}$$

Using this relationship, it is shown in many texts that the Legendre polynomials satisfy the integral equation

$$\oint d\Omega \, P_n^m P_{n'}^{m'} = \frac{2(n + |m|)! \, \delta_{n,n'}}{(2n + 1)(n - |m|)!}, \tag{7A-2}$$

and, of course, the circular functions satisfy the integral equation

$$\int_0^{2\pi} e^{i(m-m')\phi} \, d\phi = 2\pi \, \delta_{m,m'} . \tag{7A-3}$$

The combination of Eqs. (5-32), (7A-2), and (7A-3) shows that the spherical harmonics satisfy the condition

$$\oint d\Omega \, Y_n^m Y_{n'}^{m'*} = \delta_{m,m'} \, \delta_{n,n'} . \tag{7A-4}$$

In Chapter 7, we used the following orthogonality conditions:

$$\oint d\Omega\,(Q_n{}^m Q_{n'}^{m'}{}^* + S_n{}^m S_{n'}^{m'}{}^*) = n(n+1)\,\delta_{n,n'}\,\delta_{m,m'}\,, \tag{7A-5}$$

$$\oint d\Omega\,(S_n{}^m Q_{n'}^{m'}{}^* + S_{n'}^{m'}{}^* Q_n{}^m) = 0, \tag{7A-6}$$

$$\oint q\,d\Omega\,(S_n{}^m Q_n^{m'}{}^* + S_n^{m'}{}^* Q_n{}^m) = m\,\delta_{m,m'}\,, \tag{7A-7}$$

$$\oint q\,d\Omega\,(S_n^{m'} S_n^{m*} + Q_n^{m'} Q_n^{m*}) = 0, \tag{7A-8}$$

$$\oint (1-q^2)^{1/2}\,S_{n'}^{m'}\,Y_n{}^{m*}\,d\Omega = m\,\delta_{nn'}\,\delta_{mm'}\,, \tag{7A-9}$$

$$\oint Y_n{}^m Q_n^{m'}{}^*(1-q^2)^{1/2}\,d\Omega = 0. \tag{7A-10}$$

Equation (7A-5) can be evaluated by first noting that

$$S_n{}^m S_{n'}^m{}^* = \frac{m^2}{1-q^2}\,Y_n{}^m Y_{n'}^{m*}$$

and $[m^2/(1-q^2)]\,Y_n{}^m$ can be replaced by its equivalent from Eq. (5-15). Also,

$$Q_n{}^m Q_{n'}^{m*} = (1-q^2)\,\frac{dY_n{}^m}{dq}\,\frac{dY_{n'}^{m*}}{dq}\,,$$

and so the resulting integral is

$$\int_{-1}^{1} (Q_n{}^m Q_{n'}^m{}^* + S_n{}^m S_{n'}^{m*})\,dq = \left[(1-q^2)\,Y_{n'}^{m*}\,\frac{dY_n{}^m}{dq}\right]_{-1}^{1} + n(n+1)\int_{1}^{-1} Y_n{}^m Y_{n'}^{m'}{}^*\,dq. \tag{7A-11}$$

The first term on the right of Eq. (7A-11) is zero at the limits, and so, using Eq. (7A-4), Eq. (7A-11) becomes

$$\int_{1}^{-1} (Q_n{}^m Q_{n'}^{m'}{}^* + S_n{}^m S_{n'}^{m'}{}^*)\,dq = n(n+1)\,\delta_{m,m'}\,\delta_{n,n'}\,. \tag{7A-12}$$

Equation (7A-6) is confirmed by noting that

$$\int_{-1}^{1} dq(S_n{}^m Q_{n'}^{m*} + S_{n'}^{m*} Q_n{}^m) = -m \int_{-1}^{1} \frac{d}{dq}(Y_n{}^m Y_{n'}^{m*})\, dq$$

$$= -m Y_n{}^m Y_{n'}^m \Big|_{-1}^{1} = 0 \qquad (n = 1).$$

(7A-13)

The same technique used to evaluate Eqs. (7A-5) and (7A-6) confirms Eqs. (7A-7) and (7A-8) as well.

Equation (7A-3) shows that Eqs. (7A-9) and (7A-10) are both equal to zero unless $m = m'$. Using the definition of Q_n^{m*}, the integrand of Eq. (7A-10) is

$$(q^2 - 1) Y_n{}^m \frac{dY_n^{m'*}}{dq}, \tag{7A-14}$$

and from Eq. (7A-1) we note that

$$(q^2 - 1)\frac{dP_n{}^m}{dq} = [n(n+1)]^{-1}\Big[n(n - m + 1)P_{n+1}^m - (n+1)(n+m)P_{n-1}^m\Big]. \tag{7A-15}$$

The combination of Eqs. (7A-14), (7A-15), and (7A-2) gives Eq. (7A-10).

Finally, from the definition of $S_n{}^m$, Eq. (7A-9) can be rewritten as

$$\oint d\Omega\, m\, Y_n^{m'*} Y_n{}^m = m\, \delta_{m,m'}\, \delta_{n,n'}, \tag{7A-16}$$

where the equality follows from (7A-4).

Exercises

1. Find expressions for G_x and G_y, and show that each is equal to zero when a z-directed plane wave is incident.

2. Repeat Exercise 2-1 for L_x and L_y.

REFERENCE

Einstein, Albert, "Zur Quantentheorie der Strahlung," *Z. Physik.* **18,** 121–128 (1917). *English translation in* "The World of the Atom" (H. A. Boorse, and L. Motz, eds.), pp. 888–901. Basic Books, New York, 1966.

8
THE ELECTRON

In the previous chapters, we began from first principles and developed electromagnetic field theory, with the laws of classical mechanics for electromagnetic mass as a side result. But we do not have sufficient information or adequate techniques available to us to include particle theory on this general a basis. Rather, we can only note that particles exist. Three very important ones are electrons, protons, and neutrons. Some of their known properties are listed in Table 8-1.

TABLE 8-1
SOME PROPERTIES OF SEVERAL PARTICLES

	Electron	Proton	Neutron
Charge, coulombs	-1.602×10^{-19}	$+1.602 \times 10^{-19}$	0
Mass kg	9.108×10^{-31}	1.672×10^{-27}	1.675×10^{-27}
Magnetic moment, amp m^2	$1.0011\ e\hbar/2m$	$2.7928\ e\hbar/2M$	—
	9.284×10^{-24}	1.4104×10^{-26}	-0.9662×10^{-26}
Angular momentum $\hbar/2$, kg m^2/sec	—	0.5272×10^{-34}	—

TABLE 8-2
COMPARATIVE LENGTHS, METERS

	Electron	Proton
Lorentz diameter D_1	2.8179×10^{-15}	1.5347×10^{-18}
Diameter from the magnetic moment, D_3	7.6800×10^{-13}	1.1692×10^{-15}
Measured diameter	—	1.81×10^{-15}
Compton wavelength	2.426×10^{-12}	1.321×10^{-15}

The mass of a particle is given, in terms of electromagnetic fields, by Eq. (4-34). The fields produced by a slowly moving charge are given in Eqs. (4-8). It is apparent from those equations that the mass of a charge will be noninfinite only if the charge has a nonzero radius. If the charge resides on a spherical shell, then the diameter D_1 of the shell is

$$D_1 = \frac{\mu q^2}{4\pi M}, \tag{8-1}$$

where its mass is M, as shown in Eq. (4-34). If energies exist other than the electric monopolar one, then M in Eq. (8-1) is not the total mass of the particle, but rather it is only that mass which results from the electric monopole. But a nonexplosive electron is held together by a binding force. This binding force is associated with a negative binding energy and therefore produces a diameter less than the Lorentz diameter D_1.

In addition to the scalar properties of mass and charge, the net magnetic moment $\mathbf{M}$ and angular momentum $\mathbf{L}$ have been measured for several particles. These values also lead to estimates of particles size.

By definition, the angular momentum is

$$\mathbf{L} = \int \rho_M(\mathbf{r})\mathbf{r} \times \mathbf{v}\,dV, \tag{8-2}$$

where $\rho_M(\mathbf{r})$ is the mass density at each point within the particle. If the speed of each point is c, then D_2 will be smallest if $D_2 =$

$2r \sin \theta$, i.e., if the particle is a right circular cylinder of diameter D_2. The value of D_2 is, then,

$$D_2 = \frac{2L_z}{Mc}, \tag{8-3}$$

and, furthermore, the measured value of L_z is

$$L_z = \frac{\hbar}{2} \tag{8-4}$$

for each of the three particles: electrons, protons, and neutrons. This gives an expected diameter of

$$D_2 = \frac{\hbar}{Mc} \tag{8-5}$$

for each particle.

Similarly, the magnetic moment of any charged system is

$$\mathbf{M} = \tfrac{1}{2} \int \rho(\mathbf{r})\mathbf{r} \times \mathbf{v}\, dV, \tag{8-6}$$

where $\rho(\mathbf{r})$ is the charge density at a point. If the simplest possible model of a right circular cylinder of but one sign of charge is again used, then

$$D_3 = \frac{4M_z}{qc}, \tag{8-7}$$

where M_z is the magnetic moment. Now the magnetic moments for each of the three charges are given in Table 8-1. The resulting values of D_3 are very nearly

$$D_{3\mathrm{e}} = \frac{2\hbar}{mc}, \qquad D_{3\mathrm{p}} = \frac{5.59\hbar}{Mc}. \tag{8-8}$$

D_2 and D_3 are more realistic than D_1 in the sense that neither static nuclear energy nor the kinetic energy of a radially pulsating charge, e.g., a shell of radius $R(t)$, enters the calculations for D_2 and for D_3. For example, a pulsating shell of charge has a dynamic potential $\mathbf{A} = \hat{r}A(r)$, so that the magnetic field is $\mathbf{B} = \nabla \times \mathbf{A} = 0$, and, therefore, no net power is radiated. (This corresponds to the $n = 0$ case of Chapters 5 and 6.) Such pulsations would, therefore, directly affect the total mass, but they would affect the first moments only through the effect upon the average value of the diameter.

Now size estimates of particles can be made by analyzing the results of scattering beams of particles, incident upon each other at right angles. Such measurements cannot be carried out on neutrons, since they carry no net charge, and they are more difficult on electrons than on protons, since the electrons possess an equal charge but a much smaller mass; hence they are more easily deflected, and it is more difficult to produce electron–electron impact.

Of the three particles, only protons have been measured. Results of such experiments show the proton to be a multilayer structure, that it probably consists of at least three layers, and that it probably contains both positive and negative charge. The outer diameter was determined to be about 1.81×10^{-15} m, or nearly the Compton wavelength of the proton.

Now the electron is not just a scaled version of the proton, or else the ratio of angular momentum to magnetic moment in the two particles would be the same. However, to obtain an order of magnitude to the size of the electron, we note that, from each of Eqs. (8-1), (8-5), (8-7), and (8-8), the size is inversely proportional to the mass. Therefore, one size estimate is

$$D_e = \frac{M_p}{m} D_p, \tag{8-9}$$

where M_p, D_p are the mass and diameter of the proton and m, D_e are the mass and diameter of the electron. Using the measured value of D_p and Eq. (8-9), the estimated diameter is

$$D_e \approx 3.4 \times 10^{-12} \quad \text{m}. \tag{8-10}$$

Of course Eq. (8-10) depends upon the electron being a size-scaled version of the proton, and we know from comparative magnetic moment data that this is not the case.

Next we inquire whether an electron and a proton can maintain themselves in a stable equilibrium. We first consider the two particles to be separated by a distance much greater than the electron diameter. From Eq. (6-24) with $m \ll M$, the equation of motion is

$$-\frac{2a}{3c}\dddot{\mathbf{r}} + \ddot{\mathbf{r}} + \frac{ac^2}{r^2}\hat{r} = 0, \tag{8-11}$$

where each dot signifies a time derivative.

If the orbital velocity is ω, and since $\omega a \ll c$, Eq. (8-11) can be written as

$$\ddot{\mathbf{r}} + \beta\dot{\mathbf{r}} + \frac{ac^2}{r^2}\hat{r} = 0, \tag{8-12}$$

where

$$\beta = \frac{2a\omega^2}{3c}. \tag{8-13}$$

Picking the z axis in the plane of motion of the electron,

$$\begin{aligned} \dot{\mathbf{r}} &= \dot{r}\hat{r} + r\dot{\theta}\hat{\theta}, \\ \ddot{\mathbf{r}} &= \hat{r}(\ddot{r} - r\dot{\theta}^2) + \hat{\theta}(2\dot{r}\dot{\theta} + r\ddot{\theta}), \end{aligned} \tag{8-14}$$

and Eq. (8-12) separates to the two equations

$$\ddot{r} - r\dot{\theta}^2 + \beta\dot{r} + \frac{ac^2}{r^2} = 0, \tag{8-15}$$

$$\frac{d}{dt}(r^2\theta) + \beta r\dot{\theta} = 0. \tag{8-16}$$

Now the angular momentum of the rotating electron is

$$L = mr^2\dot{\theta},$$

and so Eq. (8-16) can be written as

$$\frac{dL}{dt} + \beta L = 0,$$

and, therefore,

$$L = L_0 e^{-\beta t}. \tag{8-17}$$

In words, the angular momentum decreases exponentially toward zero.

Returning to Eq. (8-15), we can estimate the energy for the case of circular motion very easily. Then $\dot{r} = 0$, and, since r is small, Eq. (8-15) gives

$$r = \frac{L^2}{ac^2m^2}. \tag{8-18}$$

A comparison of Eqs. (8-17) and (8-18) shows that r decreases exponentially toward zero; i.e., the electron will spiral into the proton.

When the distance between the electron and proton is not small compared with particle size, then stability is possible, but the detailed stable structure will depend upon detailed structure and size of the electron. And, as we saw earlier, we have no information about it.

In the absence of such information, we ask instead, what type of charge distributions and motions can exist in dynamic equilibrium which do not radiate energy permanently away from the local region containing the charge? We have already seen two examples: (1) a radially pulsating sphere of charge, and (2) a rotating ring of charge. We seek a more general method of finding possible stable but dynamic charge configurations.

To seek an answer, let us begin with the current density $\mathbf{J}(t, \mathbf{r})$ that is confined within a localized region and that is periodic with period $T = 2\pi/\omega$. The current density will have a Fourier transform given by

$$\mathbf{J}(t, \mathbf{r}) = \left[\frac{1}{2\pi}\right]^3 \sum_{l=-\infty}^{\infty} e^{i\omega_l t} \int dV_k \, \mathbf{J}(l, \mathbf{k}_l) \exp(-i\mathbf{k}_l \cdot \mathbf{r}), \tag{8-19}$$

where $\omega_l = l\omega$ and $\mathbf{k}_l$ is the wave number corresponding to the frequency ω_l. Since the current is a real quantity, it follows that

$$\mathbf{J}(l, \mathbf{k}_l) = \mathbf{J}^*(-l, -\mathbf{k}_l). \tag{8-20}$$

The use of Eqs. (2-2) and (8-19) shows the charge density to be

$$\rho(t, \mathbf{r}) = \left[\frac{1}{2\pi}\right]^3 \sum_{l=-\infty}^{\infty} e^{i\omega_l t} \int dV_k \frac{\mathbf{k} \times \mathbf{J}(l, \mathbf{k}_l)}{\omega_l} \exp(-i\mathbf{k}_l \cdot \mathbf{r}). \tag{8-21}$$

Now the vector potential $\mathbf{A}(t, \mathbf{r})$ follows from a knowledge of the current density distribution in space and time. Let $\mathbf{r}$ be the distance from an origin to the field point where $\mathbf{A}$ is to be evaluated, let $\mathbf{r}'$ be the position of the current density as measured from the same origin, and let R be the distance from the source to field points, or

$$\mathbf{R} = \mathbf{r} - \mathbf{r}'. \tag{8-22}$$

The vector potential is given by

$$\mathbf{A}(t, \mathbf{r}) = \frac{\mu}{4\pi} \int dV' \frac{\mathbf{J}[t - (R/c'), \mathbf{r}']}{R}. \tag{8-23}$$

We next assume that $r \gg r'$, so that

$$R = r - \hat{r} \cdot \mathbf{r}', \tag{8-24}$$

and then combine Eqs. (8-19), (8-23), and (8-24) to obtain

$$\mathbf{A}(t, \mathbf{r}) = \frac{\mu}{4\pi}\left[\frac{1}{2\pi}\right]^3 \sum_{l=-\infty}^{\infty} \int dV_k\, \mathbf{J}(l, \mathbf{k}_l)$$

$$\cdot \exp i\left(\omega_l t - \frac{\omega_l r}{c}\right) \int \frac{dV'}{R} \exp\left[-i\left(\mathbf{k}_l - \frac{\omega_l r}{c}\right)\right]. \tag{8-25}$$

Equation (8-25) integrates to give

$$\mathbf{A}(t, \mathbf{r}) = \frac{\mu}{4\pi r} \sum_{l=-\infty}^{\infty} \mathbf{J}(l, \mathbf{k}_l) \exp i(\omega_l t - \mathbf{k}_l \cdot \mathbf{r}), \tag{8-26}$$

where

$$\mathbf{k}_l = \frac{\omega_l}{c} \hat{r}. \tag{8-27}$$

Now the combination of Eqs. (2-12), (4-4), and (8-26) shows the field vectors to be

$$\mathbf{E}_l = -i\omega_l[\mathbf{A}_l + k(\mathbf{k}_l \cdot \mathbf{A}_l)],$$
$$\mathbf{H}_m = -\frac{i}{\mu}\mathbf{k}_m \times \mathbf{A}_m, \tag{8-28}$$

and the combination of Eqs. (4-33) and (8-28) shows the net power radiated per set of modes l and m to be

$$P_{lm} = \frac{\omega_l \omega_m}{2\eta}[A_{l\theta}A^*_{m\theta} + A_{l\phi}A^*_{m\phi}]. \tag{8-29}$$

The combination of Eqs. (8-26) and (8-29) shows the net radiated power, in terms of the current density, to be

$$P = \frac{1}{2\eta}\left[\frac{\mu}{4\pi}\right]^2 \sum_{l=-\infty}^{\infty} \sum_{n=-\infty}^{\infty} \omega_l \omega_{l-n} \exp i(\omega_n t - \mathbf{k}_n \cdot \mathbf{r})$$
$$\cdot \int d\Omega\, [\mathbf{J}_l(l, \mathbf{k}_l) \cdot (II - \hat{r}\hat{r}) \cdot \mathbf{J}^*_{l-n}(l-n, \mathbf{k}_{l-n})], \tag{8-30}$$

where II is the idem factor

$$II = \hat{r}\hat{r} + \hat{\theta}\hat{\theta} + \hat{\phi}\hat{\phi}. \tag{8-31}$$

From Eq. (8-30), if $\mathbf{J}(l, \mathbf{k}_l) = 0$ or if it is everywhere radially directed, there will be no net radiated power.

Next let us put $\boldsymbol{\xi}(t)$ equal to the position of the center of charge of a given rigid charge distribution, and the position of a given point within it, $\mathbf{r}''$, is

$$\mathbf{r}'' = \mathbf{r}' - \boldsymbol{\xi}(t).$$

$\boldsymbol{\xi}(t)$ is a periodic function of time. $\rho(\mathbf{r}'')$ is taken to be zero for all values of $r'' > b$, where b is a fixed, small constant. Putting $\mathbf{v} = \dot{\boldsymbol{\xi}}$, the current density under these conditions is

$$\mathbf{J}(\mathbf{r}'') = \mathbf{v}\rho(\mathbf{r}''). \tag{8-32}$$

Using Eq. (8-32) and the inverse Fourier transform from Eq.

(8-19), the transformed current density that appears in the power flow equation, Eq. (8-30), is

$$\mathbf{J}(l, \mathbf{k}_l) = T^{-1}\cdot\int_0^T dt \exp[-i(\omega_l t + \mathbf{k}_l\cdot\boldsymbol{\xi})]\mathbf{v}(t)$$
$$\cdot\int dV' \rho(\mathbf{r}')\exp(i\mathbf{k}_l\cdot\mathbf{r}'). \quad (8\text{-}33)$$

It is a sufficient but not necessary condition for the radiated power to be zero if $\mathbf{J}(l, \mathbf{k}_l) = 0$, or, from Eq. (8-33), if $I(\mathbf{k}_l)$, where

$$I(\mathbf{k}_l) = \int dV\rho(\mathbf{r}) \exp(i\mathbf{k}_l\cdot\mathbf{r}), \quad (8\text{-}34)$$

is equal to zero. That is, it is sufficient to assure stability to require that the Fourier transform of the charge density be zero. We next examine some special cases for which this condition is met:

Spherical shell. For a spherical shell, $\rho(\mathbf{r}) = \rho(r)$, and Eq. (8-34) becomes

$$I(k_l) = 2\pi \int_0^\pi \sin\theta\, d\theta \int_0^b r^2\, dr\, \rho(r) \exp(ikr\cos\theta), \quad (8\text{-}35)$$

where θ is the angle between $\mathbf{k}$ and r. Equation (8-35) can be written as

$$I(\mathbf{k}) = \frac{4\pi}{k}\int_0^b r\, dr\, \rho(r) \sin kr. \quad (8\text{-}36)$$

Now if q is the charge located on a spherical shell of radius $r < b$, then

$$\rho(r) = \frac{q\delta(\mathbf{r}_0 - \mathbf{r})}{4\pi r^2}. \quad (8\text{-}37)$$

The combination of Eqs. (8-36) and (8-37) shows that

$$I(\mathbf{k}_l) = q\frac{\sin kr_0}{kr_0}. \quad (8\text{-}38)$$

Equation (8-38) is 0 if $k_l r = l\pi$, where l is a nonzero integer. Putting $k_l = \omega_l/c = 2\pi/cT_l$, where T is the period of oscillation, then

$$r_0 = \frac{lcT_l}{2} \quad (8\text{-}39)$$

is a sufficient condition to assure the absence of far-field radiation.

If the shell were to rotate about an axis with period T_s, with peripheral speed $\bar{v}$, then

$$T_s\bar{v} = 2\pi r = l\pi cT$$

and

$$\frac{T}{T_s} = \frac{\bar{v}}{l\pi c}. \tag{8-40}$$

From Eq. (8-40), since l is an integer of magnitude greater than zero, the period of stable oscillation T will be less than the period of revolution T_s unless the peripheral speed $\bar{v}$ exceeds c. Therefore, either the sphere can only jiggle and cannot rotate, or it rotates with speed $\bar{v} > c$.

Other contiguous shells can be brought into the system and the results added to obtain a similar conclusion for a continuous charge distribution.

Spin. Next let us add to the current density of Eq. (8-32) a term

$$\mathbf{J}_s(\mathbf{r}') = (\Omega \times \mathbf{r}')\rho(\mathbf{r}'), \tag{8-41}$$

where $\rho(\mathbf{r}')$ is an electric charge density and, as before, the charge density is 0 for all values of r greater than b. The charge distribution is a rigid one, and it rotates with angular velocity Ω. As before, we impose the sufficient condition for stability $\mathbf{J}_s(l, \mathbf{k}_l) = 0$, or

$$\mathbf{J}_s(\mathbf{k}_l) = T^{-1}\int_0^T dt \exp[-i(\omega_l t + \mathbf{k}_l \cdot \xi)]\, \mathbf{\Omega} \times \nabla_k \int dV \rho(r) \cdot \exp(i\mathbf{k} \cdot \mathbf{r}) \tag{8-42}$$

and for $\mathbf{J}_s(\mathbf{k})$ to be 0, from Eq. (8-42), it is sufficient to put $\mathbf{I}(\mathbf{k}) = 0$, where

$$\mathbf{I}(\mathbf{k}) = \nabla_k \int dV\rho(r) \exp(ikr\cos\theta), \tag{8-43}$$

and where ∇_k signifies a gradient in k-space.

Equation (8-43) will equal zero if

$$\int_0^b r\,dr\,\rho(r)[kr\cos kr - \sin kr] = 0, \tag{8-44}$$

and Eq. (8-44) will be satisfied if $\rho(r) = B/r$, where B is a constant, and if $b = lcT$ as before.

Now we see that a charge distribution can be dynamic but still stable if

(1) it undergoes radial pulsations,

(2) it is represented by a ring of charge,

(3) it is a shell vibrating at the appropriate frequency, or if the peripheral speed can exceed *c*, or

(4) it contains internal spin but has an appropriate charge distribution.

REFERENCES

Akerlof, C. W., Hieber, R. H., Krisch, A. D., Edwards, K. W., Ratner, L. G., and Ruddick, K., "Proton–proton elastic scattering at 90° and structure within the proton," *Phys. Rev. Letters* **17,** 1105–1109 (1966).

Goedecke, G. H., "Classically radiationless motions and possible implications for quantum theory," *Phys. Rev.* **135,** 281–288 (1964).

9

AN ERGODIC EQUATION

In the previous chapter, we found some dynamic charge distributions that were in dynamic equilibrium with the field—not what systems were in local electrostatic equilibrium. Before that question can be answered, we must know more about the electron itself. At the present time, the solution of the total problem of an electron in motion about its nucleus is unknown. For that reason, and in place of a solution, we resort to other methods. One method is akin to the methods of thermodynamics in that we establish relationships between unknown quantities in the absence of complete information about the system. The other method treats the position and momentum of the electron as a random variable and then analyzes the problem as a statistical one.

Referring back to the field equations (5-33), we see that the electron can be held in equilibrium only if it generates fields for which each of Eqs. (7-7) is equal to zero. Also, the electron cannot be electrostatically stable if the field coefficients are all zero. And with nonzero coefficients the accelerating electron will exchange energy in a dynamic equilibrium with its own reactive field, just as it does during the emission and absorption processes. An example of a possible set of field coefficients that provide an appropriate equilibrium is $a_m{}^1 = a_m{}^2$, $b_m{}^1 = b_m{}^2$, and $a_m{}^1 = ib_m{}^1$.

We begin by making the simplifying assumption that the electron contains a negative charge confined within a region small compared with the total extent of the region containing the electron itself. It is in dynamic equilibrium about an attractive electrostatic potential, and, as a result of that equilibrium, there exists throughout the space around the attractive potential a time-average charge density at each point. Also, since each portion of the charge is in motion, a value of root-mean-square current density will be associated with each point.

We let $\rho(\mathbf{r})$ be the time-averaged charge density at each point. It is convenient to let $\rho(\mathbf{g})$ be the average fraction of the time the electron charge has momentum between $\mathbf{g}$ and $\mathbf{g}+d\mathbf{g}$. The electrostatic potential from any real source is continuous through all orders, and so we assume that the resulting charge densities are also continuous through all orders.

Since $\rho(\mathbf{r})$ and $\rho(\mathbf{g})$ are both negative-real functions, we can, without loss of generality, introduce

$$\begin{aligned} U(\mathbf{r})U^*(\mathbf{r}) &= -\,e^{-1}\rho(\mathbf{r}), \\ \phi(\mathbf{g})\phi^*(\mathbf{g}) &= -\,e^{-1}\rho(\mathbf{g}), \end{aligned} \tag{9-1}$$

and deal with the functions U and ϕ. They can be either positive or negative. Now U and ϕ, as defined, are different methods of describing the same phenomenon, namely, the charge density in the vicinity of the attractive potential. As such, they are functionally interdependent. Lacking detailed information, we seek those functions $U(\mathbf{r})$ and $\phi(\mathbf{g})$ that are linearly dependent, i.e., we seek functions that satisfy the relationships

$$\begin{aligned} U(\mathbf{r}) &= L[\phi(\mathbf{g})], \\ \phi(\mathbf{g}) &= L^{-1}[U(\mathbf{r})], \end{aligned} \tag{9-2}$$

where L represents a linear operator, and L^{-1} represents its inverse.

For the linear transform, we choose, just as we did earlier for electrodynamics, a Fourier integral transformation

$$\begin{aligned} U(\mathbf{r}) &= K \int \phi(\mathbf{g}) \exp\left[\frac{\mathbf{r}\cdot\mathbf{g}}{i\hbar}\right] dV_g, \\ \phi(\mathbf{g}) &= K \int U(\mathbf{r}) \exp\left[\frac{i\mathbf{r}\cdot\mathbf{g}}{\hbar}\right] dV, \end{aligned} \tag{9-3}$$

where K and $\hbar$ are constants to be determined, and where $dV_g = dg_x\, dg_y\, dg_z$.

From the definitions of $U(\mathbf{r})$ and $\phi(\mathbf{g})$, it follows that

$$\begin{aligned} \int U^*(\mathbf{r})U(\mathbf{r})\, dV &= 1, \\ \int \phi^*(\mathbf{g})\phi(\mathbf{g})\, dV_g &= 1. \end{aligned} \tag{9-4}$$

Substitution of one of Eqs. (9-4) into the other by means of Eq. (9-3) shows, after carrying out the integration, that

$$K^2 = (2\pi\hbar)^{-3}.$$

By use of Eqs. (9-4), we see that the average energy W of the charge is

$$W = (2m)^{-1} \int \phi(\mathbf{g})\phi^*(\mathbf{g})g^2\, dV_g + \int U^*(\mathbf{r})U(\mathbf{r})I(\mathbf{r})\, dV \tag{9-5}$$

if $v \ll c$, where the first term is the kinetic energy and the second term is the potential energy. The asterisk represents the complex conjugate.

If one of the functions, say $U(\mathbf{r})$, were known, then the potential energy could be calculated in Eq. (9-5), then the function $\phi(\mathbf{g})$ found by using Eq. (9-3), and then, finally, the kinetic energy found. It is more convenient, however, to proceed by another route. $g^2 = g_x^{\ 2} + g_y^{\ 2} + g_z^{\ 2}$, and we choose to illustrate the method by eliminating $g_x^{\ 2}$ from the integrand of Eq. (9-5). We do this by writing

$$\int \phi^*(\mathbf{g})\phi(\mathbf{g})g_x^{\ 2}\, dg_x = [(2\pi\hbar)^{1/2}]^{-1} \int \phi^*(\mathbf{g})g_x^{\ 2}\, dg_x \cdot \int U(\mathbf{r}) \exp\left(\frac{ixg_x}{\hbar}\right) dx,$$

where the integral of $\phi(\mathbf{g})$ over g_x has been replaced by its equivalent in Eq. (9-3). We can integrate over x once, by parts, to obtain

$$[(2\pi\hbar)^{1/2}]^{-1} \int \phi^*(\mathbf{g})g_x \frac{\hbar}{i}\, dg_x \left[U(\mathbf{r}) \exp\left(\frac{ig_x x}{\hbar}\right) - \int \frac{\partial U(\mathbf{r})}{\partial x} \exp\left(\frac{ig_x x}{\hbar}\right) dx \right]_{-\infty}^{\infty}$$

The energy calculated by Eq. (9-5) will be finite only if $U(\mathbf{r})$ and all of its derivatives approach zero as $\mathbf{r}$ increases without limit. Since the first term in the brackets must therefore be zero, the integral can be written as

$$[(2\pi\hbar)^{1/2}]^{-1}\, i\hbar \int_{-\infty}^{\infty} \phi^*(\mathbf{g}) g_x \, dg_x \int_{-\infty}^{\infty} \frac{\partial U(\mathbf{r})}{\partial x} \exp\left(\frac{i g_x x}{\hbar}\right).$$

If we again integrate by parts and impose the conditions at infinity, this expression is equal to

$$\frac{-\hbar^2}{(2\pi\hbar)^{1/2}} \int_{-\infty}^{\infty} \phi^*(\mathbf{g}) \, dg_x \int_{-\infty}^{\infty} \frac{\partial^2 U(\mathbf{r})}{\partial x^2} \exp\left(\frac{i g_x x}{\hbar}\right) dx.$$

Since these are definite integrals, we are justified in reversing the order of integration to

$$-\hbar^2 \int \frac{\partial^2 U(\mathbf{r})}{\partial x^2} \, dx \, [(2\pi\hbar)^{1/2}]^{-1} \int \phi^*(\mathbf{g}) \exp\left(\frac{i g_x x}{\hbar}\right) dg_x.$$

Finally, substituting the complex conjugate of the first of Eqs. (9-3) gives

$$\int_{-\infty}^{\infty} g_x^2 \phi^*(\mathbf{g})\phi(\mathbf{g}) \, dg_x = -\hbar^2 \int_{-\infty}^{\infty} \frac{\partial^2 U(\mathbf{r})}{\partial x^2} U^*(\mathbf{r}) \, dx.$$

The y and z components follow similarly, with the result that Eq. (9-5) can be rewritten for a general potential function $I(\mathbf{r})$ in the form

$$W = -\frac{\hbar^2}{2m} \int_{-\infty}^{\infty} U^*(\mathbf{r}) \, \nabla^2 U(\mathbf{r}) \, dV + \int_{-\infty}^{\infty} I(\mathbf{r}) U^*(\mathbf{r}) U(\mathbf{r}) \, dV,$$

and, by the first of Eqs. (9-1),

$$\int U^*(\mathbf{r}) \, dV \left[-\frac{\hbar^2}{2m} \nabla^2 U(\mathbf{r}) + [I(\mathbf{r}) - W] U(\mathbf{r}) \right] = 0. \qquad (9\text{-}6)$$

Equation (9-6) is an integrodifferential equation. It is sufficient but not necessary to require that the integrand itself be zero. Therefore, we satisfy all requirements if we put

$$-\frac{\hbar^2}{2m} \nabla^2 U(\mathbf{r}) + [I(\mathbf{r}) - W] U(\mathbf{r}) = 0. \qquad (9\text{-}7)$$

Equation (9-7) is the Schrödinger equation. The function $U(\mathbf{r})$ is the wave function. The ergodic charge density at each point is, of course, $-eU^*U$.

Next we turn to a method for obtaining Eq. (9-7) which is quite different from the preceding one. In this method, we seek, as before, a partial differential equation describing the time-average charge density $\rho(\mathbf{r})$, at each point in space, due to the motion of the charge. We consider the driving reactive force field as a random one, and we are then justified in using the central limit theorem of probability theory: the probability that a free particle will undergo a change in position $\mathbf{r}'$ during the time interval t' is

$$p(t', \mathbf{r}') = [(2\pi Dt')^{3/2}]^{-1} \exp\left[-\frac{r'^2}{2Dt'}\right], \tag{9-8}$$

where D is a constant dependent upon the characteristics of the driving field and upon the parameters of the charge. If the charge has initial velocity $\mathbf{v}_0$, then

$$p(t', \mathbf{r}') = [(2\pi Dt')^{3/2}]^{-1} \exp\left[-\frac{(\mathbf{r}' - t'\mathbf{v}_0)^2}{2Dt'}\right]. \tag{9-9}$$

Note that, for cases of interest, the velocity of the charged particle at a particular position and time will depend upon position and also, in the general nonergodic case, upon time. If we let this velocity be $\mathbf{v}'(t, \mathbf{r})$, then in the absence of acceleration the probability distribution becomes

$$P(t', \mathbf{r}'; t, \mathbf{r}) = [(2\pi Dt')^{3/2}]^{-1} \exp\left[-\frac{[\mathbf{r}' - t'\mathbf{v}'(t, \mathbf{r})]^2}{2Dt'}\right], \tag{9-10}$$

where P depends explicitly on the variables $(t, \mathbf{r})$ through $\mathbf{v}'(t, \mathbf{r})$. Since P is a probability, it satisfies the relationship

$$\int P(t', \mathbf{r}'; t, \mathbf{r})\, dV' = 1. \tag{9-11}$$

That is, the probability of the charge being somewhere at time t' is unity. We define the velocity at position $\mathbf{r}$ and time t as

$$\mathbf{v}(t, \mathbf{r}) = \lim_{t' \to 0} (t')^{-1} \int dV'\, \mathbf{r}' P(t', \mathbf{r}'; t, \mathbf{r}), \tag{9-12}$$

where the integral is over all possible positions $\mathbf{r}'$. The integral relationship is

$$\int r'^2 p(t', \mathbf{r}')\, dV' = 3Dt',$$

and, therefore, through first order in t',

$$\int r'^2 P(t', \mathbf{r}'; t, \mathbf{r})\, dV' = 3Dt' + O(t'^2), \tag{9-13}$$

where $O(t'^2)$ indicates terms on the order of t'^2. We now consider ρ a function of both time and position; i.e., $\rho = \rho(t, \mathbf{r})$. From the form of Eq. (9-10), it follows that the function $\rho(t + t', \mathbf{r})$ is

$$\rho(t + t', \mathbf{r}) = \int \rho(t, \mathbf{r} - \mathbf{r}')P(t', \mathbf{r}'; t, \mathbf{r} - \mathbf{r}')\, dV'. \tag{9-14}$$

We can obtain the ergodic equation by writing (9-14) as a differential equation. For this purpose, it is convenient to subject both sides to a Taylor series expansion, keeping terms first order in t'. The result is

$$\begin{aligned}\rho(t, \mathbf{r}) + t' \frac{\partial \rho}{\partial t} &= \int dV' \rho(t, \mathbf{r}) P(t', \mathbf{r}'; t, \mathbf{r}) \\ &\quad - \int x_i' \frac{\partial}{\partial x_i} [\rho(t, \mathbf{r}) P(t', \mathbf{r}'; t, \mathbf{r})]\, dV' \\ &\quad + \tfrac{1}{2} \int x_i' x_j' \frac{\partial}{\partial x_i} \frac{\partial}{\partial x_j} [\rho(t, \mathbf{r}) Pt', \mathbf{r}'; t, \mathbf{r})]\, dV' \\ &\quad - \cdots.\end{aligned}$$

Indices appearing twice in a single term are summed between 1 and 3. Since the integration is over the primed coordinates, $\rho(t, \mathbf{r})$ can be removed from the integration, and since $\mathbf{r}'$ is independent of $\mathbf{r}$, the x_i' can be incorporated into the partial derivative in the second integral. By making these changes, reversing the order of differentiation and integration, and writing P for $P(t', \mathbf{r}'; t, \mathbf{r})$, the equation can be written as

$$\begin{aligned}\rho + t' \frac{\partial \rho}{\rho t} &= \rho \int P dV' - \frac{\partial}{\partial x_i} [\int x_i' P dV'] \\ &\quad + \frac{1}{2} \frac{\partial}{\partial x_i} \frac{\partial}{\partial x_j} [\int x_i' x_j' P dV'] - \cdots,\end{aligned}$$

and, if Eqs. (9-11), (9-12), and (9-13) are applied through first order in t', then

$$\rho + t'\frac{\partial \rho}{\partial t} = \rho - t'\nabla \cdot (\rho \mathbf{v}) + \tfrac{1}{2}t' D\,\nabla^2 \rho$$

or

$$\frac{\partial \rho}{\partial t} = -\nabla \cdot (\rho \mathbf{v}) + \tfrac{1}{2} D\,\nabla^2 \rho. \tag{9-15}$$

Since for the ergodic case all time partials applied to averaged quantities are zero, $\nabla \cdot (\rho \mathbf{v}) = \frac{1}{2} D\,\nabla^2 \rho$. Integrating this expression once yields, within an additive constant, the average velocity:

$$\mathbf{v} = \frac{D}{2\rho}\nabla \rho. \tag{9-16}$$

This velocity arises from the stochastic reactive driving field, which, on the average, just cancels that due to the static force field.

The energy of the charge is given by

$$W = q^{-1} \int \rho(\mathbf{r})\left[\frac{m}{2}v^2 + q\phi(\mathbf{r})\right] dV. \tag{9-17}$$

Substituting the ergodic values of $\rho(\mathbf{r})$ gives

$$\int \left[\frac{m}{2}\rho v^2 + q\rho\phi - W\rho\right] dV = 0. \tag{9-18}$$

If, in turn, Eq. (9-16) is substituted into Eq. (9-18),

$$\int \left[\frac{mD^2}{8\rho}\frac{\partial \rho}{\partial x_j}\frac{\partial \rho}{\partial x_j} + q\rho\phi - W\rho\right] dV = 0.$$

We note that

$$\frac{1}{4}\frac{\partial \rho}{\rho \partial x_j}\frac{\partial \rho}{\partial x_j} = \frac{1}{2}\frac{\partial^2 \rho}{x_j x_j} - \rho^{1/2}\frac{\partial^2 \rho^{1/2}}{\partial x_j\,\partial x_j}$$

or

$$\int dV \left[\frac{mD^2}{4}\frac{\partial^2 \rho}{\partial x_j\,\partial x_j} - \frac{mD^2 \rho^{1/2}}{2}\frac{\partial^2 \rho^{1/2}}{\partial x_j\,\partial x_j} + q\rho\phi - W\right] = 0.$$

If we introduce $U^2(\mathbf{r}) = q^{-1}\rho(\mathbf{r})$, this becomes

$$\int dV \left[\frac{mD^2}{4} \frac{\partial^2 U^2}{\partial x_j \, \partial x_j} - \frac{mD^2 U}{2} \frac{\partial^2 U}{\partial x_j \, \partial x_j} + qU^2\phi - U^2 W \right] = 0, \tag{9-19}$$

where $\int \rho(\mathbf{r}) \, dV = q$. Now by Gauss's law $\int \nabla^2 U^2 \, dV = \int \nabla U^2 \cdot d\mathbf{S}$, and, in the limit of large distances, U^2 becomes vanishingly small. Hence the first term of Eq. (9-19) is itself zero. This leaves the integrodifferential equation

$$\int dV \, U \left[-\frac{mD^2}{2} \nabla^2 U + (q\phi - WU) \right] = 0. \tag{9-20}$$

The requirement is that the integral be equal to zero. This requirement can be more than sufficiently met by demanding that the integrand itself be everywhere equal to zero. Let us substitute a new constant

$$\hbar = mD. \tag{9-21}$$

The combination of Eqs. (9-20) and (9-21) gives

$$-\frac{\hbar^2}{2m} \nabla^2 U + (I - W)U = 0. \tag{9-22}$$

This is, once again, the Schrödinger time-independent equation, and is the differential equation that the ergodic charge density must satisfy. Since it is based on a time average, it is valid only for times extending over many changes in the driving force, i.e., over times long enough for the force to undergo many changes and for average charge density to approach the ergodic value arbitrarily closely.

From the theory of random walk, the diffusion constant D is also given by

$$D = \frac{\omega L^2}{4\pi}, \tag{9-23}$$

where ω is the mean radian frequency of the disturbing field, and L is the mean distance the electron travels between disturbances. To establish an order of magnitude for L, let us put

$$L = \frac{2\pi c}{\omega}. \tag{9-24}$$

The combination of Eqs. (9-21), (9-23), and (9-24) shows L to be

$$L \approx \frac{2\hbar}{mc}. \tag{9-25}$$

We note that Eq. (9-25) is the same as the electron size as calculated by using magnetic moment considerations, Eq. (8-8). The elimination of L from Eqs. (9-24) and (9-25) shows

$$\hbar\omega \approx mc^2. \tag{9-26}$$

REFERENCES

Grimes, D. M., "Point charge in a box; a new approach to an old problem," *Proc. IEEE* **53,** 737 (1965).

Kershaw, David, "Theory of hidden variables," *Phys. Rev.* **136B**, 1850–1856 (1964).

10

PROPERTIES OF THE WAVE FUNCTION

Detailed solution of the Schrödinger equation depend upon the function $I(\mathbf{r})$ and the value of the constant W. We can learn much about the solutions, however, without knowing what they are. To do so, we return to the method of partial integration used in Chapter 9 for obtaining average values of momentum in terms of $U(\mathbf{r})$; the result is

$$\int \phi^*(\mathbf{g}) g_i^n g_j^m \phi(\mathbf{g})\, dV_g = (i\hbar)^{n+m} \int U^*(\mathbf{r}) \frac{\partial^{n+m}}{\partial x_i^n \partial x_j^m} U(\mathbf{r})\, dV. \quad (10\text{-}1)$$

n and m are integers, and the subscripts i and j indicate the ith and jth directions, respectively. Equation (10-1) expresses ergodic values of momentum in terms of the coordinate-space function $U(\mathbf{r})$. Of course, these values have significance only when integrated over all charge-bearing regions or over times long enough for average values to be obtained.

We have constructed the wave functions as complex functions for ease of calculation and, in anticipation of the next section, for ease in going to the time-dependent case. But any result that corresponds to a physically measurable quantity must be real. Let

$$\int U^*{}_a(\mathbf{r}) \hat{O}(\mathbf{r}, \mathbf{g}) U_b(\mathbf{r})\, dV \quad (10\text{-}2)$$

represent a physically measurable quantity of the charged system, where $U_a^*(\mathbf{r})$ and $U_b(\mathbf{r})$ are wave functions and $\hat{O}(\mathbf{r}, \mathbf{g})$ represents a function of position and momentum, if expansions in momentum are written according to Eq. (10-1). Then this integral must be real, that is, must equal its own complex conjugate; and after each term in Eq. (10-1) is integrated by parts as many times as it contains powers of **g**,

$$\int U_a^*(\mathbf{r})\hat{O}(\mathbf{r}, \mathbf{g})U_b(\mathbf{r})\,dV = \int U_b^*(\mathbf{r})\hat{O}(\mathbf{r}, \mathbf{g})U_a(\mathbf{r})\,dV. \qquad (10\text{-}3)$$

Operators $\hat{O}(\mathbf{r}, \mathbf{g})$ that satisfy Eq. (10-3) are Hermitian operators. Every operator that produces a real average value, as these do, corresponds to a physically measurable quantity. Every physically measurable quantity can be formulated as a Hermitian operator, that is, will satisfy Eq. (10-3). A particularly useful example is $\hat{O}(\mathbf{r}, \mathbf{g}) = (2m)^{-1}\ g^2 + I(\mathbf{r})$, which is the total energy, usually called the Hamiltonian. In operator notation, this is

$$\hat{O}(\mathbf{r}, \mathbf{g}) = \left(-\frac{\hbar^2}{2m}\nabla^2 + I\right) = \hat{H}, \qquad (10\text{-}4)$$

so that the Schrödinger equation becomes

$$\hat{H}U = WU. \qquad (10\text{-}5)$$

Now let $U_a(\mathbf{r})$ and W_a be a member of a set of N possible solutions of the Schrödinger equation. The sum over all solutions is also a solution; that is,

$$U(\mathbf{r}) = \sum_{a=1}^{N} C_a U_a(\mathbf{r}) \qquad (10\text{-}6)$$

represents the total solution to the set of equations. Now let us use Eq. (10-5) to construct the integral expressions

$$\begin{aligned} \int U_b^*\hat{H}U_a\,dV &= W_a \int U_b^* U_a\,dV, \\ \int U_a^*\hat{H}U_b\,dV &= W_b \int U_a^* U_b\,dV, \end{aligned} \qquad (10\text{-}7)$$

where a and b are members of the set of solutions of Eq. (10-5). The combination of Eqs. (10-3) and (10-7) shows the left sides of Eqs. (10-7) to be equal and, since energies are real, that

$$(W_a - W_b)\int U_b^* U_a\,dV = 0. \qquad (10\text{-}8)$$

From Eq. (10-8), we see that, for $W_a \neq W_b$, the integral is zero and, therefore, using Eqs. (9-4), that

$$\int U_b{}^* U_a \, dV = \delta_{ab}, \tag{10-9}$$

where $\delta_{ab} = 1$ if $a = b$ and is 0 otherwise. That is, the wave functions form an orthogonal set. By definition of the wave function $U(r)$, then, the average charge density would be $-eU_a{}^*U_a$ if U_a were the only solution. It follows from combining Eqs. (10-5) and (10-6) that

$$\sum_{a=1}^{N} C_a{}^* C_a = 1, \tag{10-10}$$

and, therefore, that $C_a{}^*C_a$ is the fraction of time during which the charge is described by the wave function U_a; that is, it is the probability that the charge density is $-eU_a{}^*U_a$.

The coefficients C_a cannot be found, of course, from the differential equation, but, in the absence of disturbing influences, the state with the lowest value of energy, say W_i, will have a probability of 1, or $C_i = 1$ and $C_a = 0$, for $a \neq i$.

Another property of the wave functions can be found by examining the dipole moment of the charge distribution,

$$\mathbf{P} = -e \int U^*(\mathbf{r}) \mathbf{r} U(\mathbf{r}) \, dV. \tag{10-11}$$

Unless the potential field $I(\mathbf{r})$ is dipolar, there is no mechanism by which a nonzero dipole moment can be produced. Equation (10-11) must, therefore, be zero for all wave functions $U(\mathbf{r})$; but this is the case only if U^*U is an even function of $\mathbf{r}$. To apply this result, we begin by noting that any function can be broken down into the sum of its even and odd components; that is, $U(\mathbf{r})$ can be written as

$$U(\mathbf{r}) = U_e(\mathbf{r}) + U_o(\mathbf{r}), \tag{10-12}$$

where the subscripts e and o represent even and odd, respectively. The product U^*U, written to incorporate Eq. (10-12), is

$$U^*(\mathbf{r})U(\mathbf{r}) = (U_e U_e{}^* + U_o U_o{}^*) + (U_e{}^* U_o + U_e U_o{}^*). \tag{10-13}$$

When the first term on the right side of Eq. (10-13) is placed in the integrand of Eq. (10-12), the result is zero. This is not true for the second term. Hence, either U_e or U_o must be zero; i.e., $U(\mathbf{r})$ is either an even function or an odd one, but is not a combination of both. $U(\mathbf{r})$ is said to have even or odd parity according to whether it is even or odd itself.

Let us next consider two separate electrostatic potential functions. Each attracts a single charge, and they are far enough apart to be independent of each other. For this case, Eq. (10-5) is

$$\begin{aligned} -\frac{\hbar}{2m}\nabla'^2 U_a'(\mathbf{r}') + [I'(\mathbf{r}') + W_a']U_a'(\mathbf{r}') &= 0, \\ -\frac{\hbar}{2m}\nabla''^2 U_b''(\mathbf{r}'') + [I''(\mathbf{r}'') + W_b'']U_b''(\mathbf{r}'') &= 0, \end{aligned} \tag{10-14}$$

where $U_a'(\mathbf{r}')$ represents a wave function that is a solution to the Schrödinger equation in which the potential is $I'(\mathbf{r}')$ and the energy is W_a', and $U_b(\mathbf{r}'')$ represents a wave function that is a solution to the Schrödinger equation in which the potential is $I''(\mathbf{r}'')$ and the energy is W_b''. Since these equations and their solutions are independent, Eq. (10-14) can be rewritten as

$$-\frac{\hbar^2}{2m}\nabla^2 U_{ab}(\mathbf{r}', \mathbf{r}'') + [I(\mathbf{r}', \mathbf{r}'') - W_{ab}]U_{ab}(\mathbf{r}', \mathbf{r}'') = 0, \tag{10-15}$$

where

$$\begin{aligned} \nabla^2 &= \nabla'^2 + \nabla''^2, \\ U_{ab}(\mathbf{r}', \mathbf{r}'') &= U_a'(\mathbf{r}')U_b''(\mathbf{r}''), \\ I(\mathbf{r}', \mathbf{r}'') &= I'(\mathbf{r}') + I''(\mathbf{r}''), \\ W_{ab} &= W_a' + W_b'', \end{aligned} \tag{10-16}$$

and the total wave function $U(\mathbf{r}', \mathbf{r}'')$ is the sum of all possible product functions $U_{ab}(\mathbf{r}', \mathbf{r}'')$, weighted according to the probability of each state; that is,

$$U(\mathbf{r}', \mathbf{r}'') = \sum_{a=1}^{N'} \sum_{b=1}^{N''} C_{ab} U_{ab}(\mathbf{r}', \mathbf{r}''), \tag{10-17}$$

where

$$C_{ab} = C_a{}'C_b{}''.$$

To find the desired property of the wave function, we next bring the two systems into proximity in a particular way. We place two identical charges in the same electrostatic potential field at positions $\mathbf{r}'$ and $\mathbf{r}''$; for the moment, we ignore interaction between them. For this case, the wave function is still expressed by Eq. (10-17), but now $U_i(\mathbf{r}')$ is identical in form with $U_i(\mathbf{r}'')$ and $N' = N''$, or

$$U(\mathbf{r}', \mathbf{r}'') = \sum_{a=1}^{N'} \sum_{b=1}^{N''} C_{ab} U_a(\mathbf{r}') U_b(\mathbf{r}''), \qquad (10\text{-}18)$$

which can be rewritten as

$$U(\mathbf{r}', \mathbf{r}'') = \sum_{a=1}^{N} \sum_{b=1}^{N} C_{ab}[U_a(\mathbf{r}')U_b(\mathbf{r}'') - U_b(r')U_a(\mathbf{r}'')]$$
$$+ D_{ab}[U_a(\mathbf{r}')U_b(\mathbf{r}'') + U_b(\mathbf{r}')U_a(\mathbf{r}'')] \qquad (10\text{-}19)$$

where $C_{ab} = \frac{1}{4}[C_{ab} - C_{ba}]$ and where $D_{ab} = \frac{1}{4}[C_{ab} + C_{ba}]$. Equation (10-19) can be written more compactly as

$$U(\mathbf{r}', \mathbf{r}'') = \sum_{a=1}^{N} \sum_{b=1}^{N} [C_{ab} U_{ab}(\mathbf{r}', \mathbf{r}'') + D_{ab} V_{ab}(\mathbf{r}', \mathbf{r}'')], \qquad (10\text{-}20)$$

where

$$\begin{aligned} U_{ab} &= U_a(\mathbf{r}')U_b(\mathbf{r}'') - U_b(\mathbf{r}')U_a(\mathbf{r}''), \\ V_{ab} &= U_a(\mathbf{r}')U_b(\mathbf{r}'') + U_b(\mathbf{r}')U_a(\mathbf{r}''), \end{aligned} \qquad (10\text{-}21)$$

and, from Eqs. (10-21),

$$\begin{aligned} U_{ab}(\mathbf{r}', \mathbf{r}') &= 0, \qquad V_{ab}(\mathbf{r}', \mathbf{r}') = 2U_a(\mathbf{r}')U_b(\mathbf{r}'), \\ U_{aa}(\mathbf{r}', \mathbf{r}'') &= 0, \qquad V_{aa}(\mathbf{r}', \mathbf{r}'') = 2U_a(\mathbf{r}')U_b(\mathbf{r}''). \end{aligned} \qquad (10\text{-}22)$$

Equation (10-20) is the final form for the model analyzed.

A question basic to our further discussion is whether the function UU^* represents the fraction of the static charge occupying differential volume dV or whether it is the time-average charge density within that volume—that average produced within each differential volume by virtue of the electron's motion through it. On one hand, if the charge distribution is a static one, then only the total energy of the electrostatic charge system need be considered. On the other hand, if the electron is localized within a volume small compared with the volume throughout which it moves, then not only must the time and spatial average of dynamic quantities satisfy the ergodic hypothesis of being equal, but also there will exist a maximum potential energy that the electron can have at any instant. We saw in Chapter 8 that the most probable electron diameter is about 3.4×10^{-12} m, and we shall see in Chapter 15 that the expected atomic radius is about 10^{-10} m. Therefore, the interpretation of UU^* as a time-average value of electronic charge density at a point, as the electron is driven by its own reactive field, seems most likely to be correct.

Next let us consider the interaction energy between two charges, that interaction energy we previously ignored. If we let $\mathscr{E}$ represent the electrostatic energy between the charges, then the interaction energy between differential volumes $dV'\,dV''$ is

$$d\mathscr{E} = \frac{e^2}{4\pi\varepsilon}\,\frac{P(\mathbf{r}', \mathbf{r}')\,dV'\,dV''}{|\mathbf{r}' - \mathbf{r}''|}, \tag{10-23}$$

where $P(\mathbf{r}', \mathbf{r}'')$ is the probability that the charges will simultaneously occupy volumes dV' and dV'', and is equal to $U^*(r', \mathbf{r}'')U(r', r'')$. Integrating Eq. (10-23) gives the total interaction energy:

$$\mathscr{E} = \frac{e^2}{4\pi\varepsilon} \sum_{p=1}^{N} \sum_{q=1}^{N} \sum_{a=1}^{N} \sum_{b=1}^{N} \int dV' \int dV''\,(\mathbf{r}' - \mathbf{r}'')^{-1} \times (C_{pq}U_{pq} + D_{pq}V_{pq})\,(C_{ab}U_{ab} + D_{ab}V_{ab})^*, \tag{10-24}$$

where C and D are the constants as defined in Eq. (10-20). Of these integrals, we concentrate on

$$\int dV' \int dV''\,(|\,\mathbf{r}' - \mathbf{r}''\,|)^{-1}\,(C_{ab}{}^*C_{ab}U_{ab}{}^*U_{ab} + D_{ab}{}^*D_{ab}V_{ab}{}^*V_{ab} + \cdots + \cdots),$$

noting from Eq. (10-22) that, when $\mathbf{r}' = \mathbf{r}''$, $U_{ab} = 0$ but $V_{ab} \neq 0$.

Turning to the energy integral of Eq. (10-24), the integrand using the antisymmetric functions is regular at all points, and the integrand using the symmetric functions is singular when $\mathbf{r}' = \mathbf{r}''$. Although, for many acceptable wave functions $U(\mathbf{r})$, the integrals themselves remain finite, the electron cannot carry out a trajectory that contains a large interaction energy, such as occurs with symmetric wave functions at $\mathbf{r}' = \mathbf{r}''$. We conclude, therefore, that only solutions for which $D_{ab} = 0$ can exist in equilibrium. For this case, the total wave function is

$$U(\mathbf{r}', \mathbf{r}'') = \sum_{a=1}^{N} \sum_{b=1}^{N} C_{ab} U_{ab}(\mathbf{r}', \mathbf{r}''). \qquad (10\text{-}25)$$

That is, if the problem is to be formulated without accounting for intercharge forces, then we must separately impose the condition that the wave function be antisymmetric with respect to an interchange of position of the charges. Combining this result with Eq. (10-23) shows that the probability of a single volume element's being doubly occupied is zero, and that the probability of the two charge densities being described by the same wave function is also zero. This important result is the Pauli exclusion principle.

We turn now to yet another general effect, the Heisenberg uncertainty principle. For this purpose, we examine $\langle x^2 \rangle$ $\langle g_x{}^2 \rangle$. As was shown in Chapter 9, the integral expression for $\langle g_x{}^2 \rangle$ is

$$\langle g_x{}^2 \rangle = -\hbar^2 \int U^*(\mathbf{r}) \frac{d^2 U(\mathbf{r})}{dx^2} \, dV.$$

When this is integrated once by parts and the limits are imposed at infinity, it becomes

$$\langle g_x{}^2 \rangle = \hbar^2 \int \frac{dU^*(\mathbf{r})}{dx} \frac{dU(\mathbf{r})}{dx} \, dV.$$

Combining this with the expression for $\langle x^2 \rangle$ gives

$$\langle x^2 \rangle \langle g_x{}^2 \rangle = \hbar^2 \int U^* U x^2 \, dV \int \frac{dU^*}{dx} \frac{dU}{dx} \, dV. \qquad (10\text{-}26)$$

Now we consider f and p to be two complex functions, say $f_1 + if_2$ and $p_1 + ip_2$, and consider the expression $4ff^*pp^* - (f^*p + fp^*)^2$. Written out in detail, this expression is

$$4(f_1^2 + f_2^2)(p_1^2 + p_2^2) - 4(f_1 p_1 + f_2 p_2)^2,$$

which simplifies to

$$4(f_1 p_2 - f_2 p_1)^2 \geqslant 0,$$

where the inequality is imposed because the left side is positive–real. This is equivalent to

$$ff^*pp^* \geqslant \tfrac{1}{4}(f^*p + fp^*)^2,$$

the Schwarz inequality. If this inequality is written in integral form with f and p complex functions of position, the result is

$$4 \int f^*(r)f(r)\,dV \int p^*(r)p(r)\,dV \geqslant \{ \int [f^*(r)p(r) + f(r)p^*(r)]\,dV\}^2. \tag{10-27}$$

To apply this inequality, we put

$$f(r) = xU(r),$$

$$p(r) = \frac{dU(r)}{dx},$$

and thus obtain

$$\int dV \left(xU^* \frac{dU}{dx} + xU \frac{dU^*}{dx} \right)^2 \leqslant 4 \int x^2 U^*\,dV \int \frac{dU}{dx}\frac{dU^*}{dx}\,dV. \tag{10-28}$$

Now we also note that

$$x \frac{d}{dx}(U^*U) = xU^* \frac{dU}{dx} + xU \frac{dU^*}{dx}, \tag{10-29}$$

and that

$$\int x \frac{d}{dx}(U^*U)\,dV = -\int U^*U\,dV. \tag{10-30}$$

Combining Eqs. (10-26), (10-28), (10-29), and (10-30) gives

$$[\langle x^2 \rangle \langle g_x \rangle^2]^{1/2} \geqslant \frac{\hbar}{2}. \tag{10-31}$$

Equation (10-31) is a quantitative statement of the Heisenberg uncertainty principle. $\langle x^2 \rangle^{1/2}$ represents the root-mean-square range of positions occupied by the electron. If in some way this range is made to decrease toward zero, the range of momentum must increase parabolically, and vice versa. The interpretation is that when the charge is subject to a reactive driving field, its coordinates can be known only statistically.

We note in passing that, by combining the exclusion principle with the Hermitian property of quantum-mechanical operators, we can obtain a result for a system containing two charges within a single electrostatic potential field. The mean-square distance between charges at positions r' and r'' is

$$\langle R^2 \rangle = \langle |\mathbf{r}' - \mathbf{r}''| \rangle^2 = \langle r'^2 \rangle + \langle r''^2 \rangle - 2\langle \mathbf{r}' \rangle \cdot \langle \mathbf{r}'' \rangle,$$

and the wave function is in the form of Eq. (10-25). We note that the properly normalized wave function for charges in states m and n is

$$(\surd 2)^{-1} [U_m(\mathbf{r}')U_n(\mathbf{r}'') - U_n(\mathbf{r}')U_m(\mathbf{r}'')],$$

so that the expectation value of R^2 for states m and n is

$$\langle R^2 \rangle = \tfrac{1}{2} \int dV' \int dV'' [U_m^*(\mathbf{r}')U_n^*(\mathbf{r}'') - U_n^*(\mathbf{r}')U_m^*(\mathbf{r}'')] \times [r'^2 + r''^2 - 2\mathbf{r}' \cdot \mathbf{r}''][U_m(\mathbf{r}')U_n(\mathbf{r}'') - U_n(\mathbf{r}')U_m(\mathbf{r}'')]. \tag{10-32}$$

Equation (10-32) integrates to

$$\langle R^2 \rangle = \langle r_m{}^2 \rangle + \langle r_n{}^2 \rangle - 2\langle \mathbf{r}_n \rangle \cdot \langle \mathbf{r}_m \rangle + 2 \int U_m^* \mathbf{r} U_n \, dV \cdot \int \mathbf{r} U_n^* U_m \, dV.$$

The last term is of particular interest. From Eq. (10-3), it can be rewritten as

$$2[\int U_m^*(\mathbf{r})\mathbf{r}U_n(\mathbf{r}) \, dV]^2 = 2(m \,|\, \mathbf{r} \,|\, n)^2 \geqslant 0, \tag{10-33}$$

so that Eq. (10-32) can be written as

$$\langle R^2 \rangle = \langle r_m{}^2 \rangle + \langle r_n{}^2 \rangle - 2\langle \mathbf{r}_m \rangle \cdot \langle \mathbf{r}_n \rangle + 2(m \,|\, \mathbf{r} \,|\, n)^2, \tag{10-34}$$

where the last term arises from the antisymmetry of the wave function. The interpretation is that the repulsive force between the charges increases the mean-square distance between them.

11

A TIME-DEPENDENT BUT ERGODIC EQUATION

The rate of change of the average value of any dynamic quantity is found by combining the time-dependent solution with the continuity equation. The average charge density created by charge q at each point is given by qU^*U and, by definition, is independent of time within periods of the length with which we have so far been concerned. A time-dependent function $\psi(t, \mathbf{r})$ can be defined, however, which is equal to

$$\psi(t, \mathbf{r}) = U(\mathbf{r})e^{i\omega t}, \tag{11-1}$$

where ω is a frequency as yet undetermined. All results that can be checked experimentally will depend upon the product $\psi^*\psi = U^*U$, which is independent of time. The time introduced in Eq. (11-1) represents not at all a true time-sequence of events, but rather a rate of change of ergodic values over much longer periods.

For purposes of calculation, we can use Eq. (10-26) to determine the momentum density at each point in space as $i\hbar U^* \nabla U$. The effective current density is, then $(iq\hbar/m)U^* \nabla U$ for a charge q with mass m. Recall that this current density must be real; by Eq. (10-26), therefore,

$$\frac{iq\hbar}{2m}(\psi^* \nabla\psi - \psi \nabla\psi^*) = \mathbf{J}_e, \tag{11-2}$$

where $\mathbf{J}_e$ is the effective current density at a point.

Now the continuity equation expresses the conservation of electric charge, and is

$$\nabla \cdot \mathbf{J} + \frac{\partial \rho}{\partial t} = 0. \tag{11-3}$$

For the ergodic case, this must apply also to the time-average values produced by a moving charge. When the appropriate values from Eqs. (9-1) and (11-2) are inserted, and the result is integrated over all space, Eq. (11-3) becomes

$$\int \left[\psi^* \left(\frac{iq\hbar}{2m} \nabla^2 \psi + q \frac{\partial \psi}{\partial t} \right) + \text{complex conjugate} \right] dV_r = 0.$$

If we multiply by $i\hbar/q$, and add and subtract $I(\mathbf{r})\psi^*\psi$, Eq. (11-3) becomes

$$\int dV_r \left[\psi^* \left(-\frac{\hbar^2}{2m} \nabla^2 \psi + I(\mathbf{r}) \psi + i\hbar \frac{\partial \psi}{\partial t} \right) - \text{complex conjugate} \right] = 0. \tag{11-4}$$

Comparing Eqs. (9-7) and (11-4) shows that the latter is real and equal to the Schrödinger equation if

$$W\psi = \frac{\hbar}{i} \frac{\partial \psi}{\partial t}, \tag{11-5}$$

where

$$-\frac{\hbar^2}{2m} \nabla^2 \psi + I\psi + i\hbar \frac{\partial \psi}{\partial t} = 0. \tag{11-6}$$

This is the Schrödinger time-dependent equation, where

$$\psi(t, \mathbf{r}) = U(\mathbf{r}) \exp \frac{iWt}{\hbar}, \tag{11-7}$$

where, clearly, the frequency is given by

$$\omega = \frac{W}{\hbar}. \tag{11-8}$$

We note that Eq. (11-8) is nearly the same as Eq. (9-26).

As stated, the time dependence described here represents a time change of average or ergodic values, and not in any sense an actual time sequence of events.

Equation (11-6) is often abbreviated as

$$\hat{H}\psi = \frac{\hbar}{i}\frac{\partial\psi}{\partial t}, \tag{11-9}$$

where $\hat{H}$ is the sum of all energy terms, written in operational form to make it usable for calculating average values in coordinate space.

As an example of the use of Eq. (11-7), let us find the rate of change $(d/dt)\,\langle \mathbf{O}(\mathbf{r}, \mathbf{g})\rangle$. This is

$$\frac{d}{dt}\langle \mathbf{O}\rangle = \frac{d}{dt}\int \psi^*\hat{\mathbf{O}}\psi\, dV = \int dV\left(\frac{\partial\psi^*}{\partial t}\,\hat{\mathbf{O}}\psi + \psi^*\hat{\mathbf{O}}\,\frac{\partial\psi}{\partial t}\right),$$

where the functions depend upon time explicitly, but $\hat{\mathbf{O}}$ does not. Substituting values from Eq. (11-6) gives

$$\frac{d}{dt}\langle \mathbf{O}(\mathbf{r}, \mathbf{g})\rangle = \frac{i}{\hbar}\int dV\,[U^*\hat{\mathbf{O}}(\hat{H}U) - (\hat{\mathbf{O}}U)\hat{H}^*U^*].$$

Since $\hat{\mathbf{O}}$ and $\hat{H}$ are Hermitian operators,

$$\frac{d}{dt}\langle \mathbf{O}(\mathbf{r}, \mathbf{g})\rangle = \frac{i}{\hbar}\int dV\, U^*[\hat{\mathbf{O}}(\hat{H}U) - \hat{H}(\hat{\mathbf{O}}U)],$$

and

$$\frac{d}{dt}\langle \mathbf{O}(\mathbf{r}, \mathbf{g})\rangle = \frac{i}{\hbar}\int dV\, U^*[\hat{\mathbf{O}}\hat{H} - \hat{H}\hat{\mathbf{O}}]U, \tag{11-10}$$

where, as always, the operators operate on everything multiplicative to their right. The commutator of $\hat{\mathbf{O}}$ is

$$\frac{i}{\hbar}[\hat{\mathbf{O}}\hat{H} - \hat{H}\hat{\mathbf{O}}].$$

So the time dependence of the expectation value of **O** is in proportion to the commutator of **O**.

For the special case of $\mathbf{O}(\mathbf{r}, \mathbf{g}) = \mathbf{r}$, it is easy to show that

$$\frac{d}{dt}\langle\mathbf{r}\rangle = m^{-1}\langle\mathbf{g}\rangle, \tag{11-11}$$

and for $\mathbf{O}(\mathbf{r}, \mathbf{g}) = \mathbf{g}$ that

$$\frac{d}{dt}\langle\mathbf{g}\rangle = -q\langle\nabla I\rangle. \tag{11-12}$$

Since the right side of Eq. (11-12) is the force $\mathbf{F}$ on the charge, combining Eqs. (11-11) and (11-12) gives

$$\langle\mathbf{F}\rangle = m\frac{d^2}{dt^2}\langle\mathbf{r}\rangle. \tag{11-13}$$

Thus, expectation values obey Newton's laws.

12
THE RADIATING ATOM

Chapters 6 and 7 contain discussions of those radiative properties of a volume which are independent of the detailed characteristics of the volume and depend only upon properties of radiation itself. Chapter 9 contains an analysis of properties of regions containing charge by analyzing the region itself, assuming that the region is driven by a reactive field, but without a net energy exchange. Now a complete solution of the problem must, of course, contain an analysis of both the radiation and its source.

Until this point, when dealing with radiation, we considered a small object to radiate, but applied no restrictions to that object other than requiring it to be capable of sustaining the flow of alternating current. We next impose a further restriction upon the radiating system: the average electron energy within time-stable atoms is the constant value W given by Eq. (9-7).

We begin by considering the power flow into and out of an object under the influence of radiant energy at two unrelated frequencies ω_p and ω. Since we are interested in power flow that is independent of system details, let us take a single lossless capacitor as a model. The voltage across the capacitor is a unique but possibly nonlinear function of the charge stored on each plate. If we let capital letters denote the undetermined parameters, the charge q stored on each plate can be expressed as the sum over all possible frequencies:

$$q(t) = \frac{1}{2} \sum_{m=-\infty}^{m=\infty} \sum_{n=-\infty}^{n=\infty} Q_{mn} \exp i(m\omega_p + n\omega)t, \tag{12-1}$$

where the Q_{mn} are time-independent and $Q_{mn} = Q^*_{-m,-n}$. Similarly, the voltage across the capacitor can be expressed as

$$v(t) = \frac{1}{2} \sum_{m=-\infty}^{m=\infty} \sum_{n=-\infty}^{n=\infty} V_{mn} \exp i(m\omega_p + n\omega)t, \tag{12-2}$$

where the parameters V_{mn} are time-independent and $V_{mn} = V^*_{-m,-n}$. Differentiating Eq. (12-1) with respect to time gives the current flowing into the capacitor:

$$i(t) = \frac{1}{2} \sum_{m=-\infty}^{m=\infty} \sum_{n=-\infty}^{n=\infty} I_{mn} \exp i(m\omega_p + n\omega)t, \tag{12-3}$$

where $I_{mn} = i(m\omega_p + n\omega)Q_{mn}$, and so $I_{mn} = I^*_{-m,-n}$.

Since the capacitor is lossless, the average power flow is zero. In terms of the preceding coefficients, the average power P_{mn} flowing at frequency $m\omega_p + n\omega$ is

$$P_{mn} = \tfrac{1}{2} \operatorname{Re} (V_{mn} I^*_{mn})$$

or

$$P_{mn} = -\tfrac{1}{2}(m\omega_p + n\omega) \operatorname{Re} (iV_{mn} Q^*_{mn}), \tag{12-4}$$

and P_{mn} summed over all frequencies is the total power flow P. It can be written as

$$P = \frac{1}{2} \sum_{m=-\infty}^{m=\infty} \sum_{n=-\infty}^{n=\infty} P_{mn} = 0. \tag{12-5}$$

To arrange Eq. (12-5) in a more useful form, we multiply both numerator and denominator by $m\omega_p + n\omega$ and separate the sums into two parts as

$$\omega_p \sum_{m=-\infty}^{m=\infty} \sum_{n=-\infty}^{n=\infty} \frac{mP_{mn}}{(m\omega_p + {}_m\omega)} + \omega \sum_{m=-\infty}^{m=\infty} \sum_{n=-\infty}^{n=\infty} \frac{nP_{mn}}{(m\omega_p + n\omega)} = 0. \tag{12-6}$$

By Eq. (12-4), the ratio $P_{mn}/(m\omega_p + n\omega)$ is equal to the product

$V_{mn}Q^*_{mn}$, which depends upon the permittivity of the capacitor, but not upon the frequency. Hence Eq. (12-6) can be written as

$$a\omega_p + b\omega = 0,$$

where a and b are constants and where ω_p and ω are unrelated frequencies. This can be satisfied only if each coefficient is zero, that is, if each sum is separately zero. They can be rewritten as

$$\sum_{m=-\infty}^{m=\infty} \sum_{n=-\infty}^{n=\infty} \frac{mP_{mn}}{m\omega_p + n\omega} = 0,$$
$$\sum_{m=-\infty}^{m=\infty} \sum_{n=-\infty}^{n=\infty} \frac{nP_{mn}}{m\omega_p + n\omega} = 0. \qquad (12\text{-}7)$$

From the relationships for V_{mn}, I_{mn}, and P_{mn}, it follows that $P_{mn} = P_{-m,-n}$ and, therefore, that the first of Eqs. (12-7) can be written as

$$\sum_{m=1}^{\infty} \sum_{n=-\infty}^{\infty} \frac{mP_{mn}}{m\omega_p + n\omega} + \sum_{m=-\infty}^{-1} \sum_{n=-\infty}^{\infty} \frac{mP_{mn}}{m\omega_p + n\omega},$$

and that the two terms are equal. A similar argument follows for the second of Eqs. (12-7) with the result

$$\sum_{m=0}^{\infty} \sum_{n=-\infty}^{\infty} \frac{mP_{mn}}{m\omega_p + n\omega} = 0,$$
$$\sum_{m=-\infty}^{\infty} \sum_{n=0}^{\infty} \frac{nP_{mn}}{m\omega_p + n\omega} = 0. \qquad (12\text{-}8)$$

These are the Manley–Rowe equations. For a linear capacitor, only Q_{10} and Q_{01} are different from zero, but for a nonlinear one other components can exist. Their magnitudes are determined by the properties of the capacitor; that is, the exact values depend upon the detailed structure of the capacitor. We now proceed to apply these equations to an atom.

Solutions of the Schrödinger equation, as discussed in Chapter 10, shows that any confined system must have discrete energy states. By Eq. (11-8) the energy of the state determines the mean frequency, $\omega_p = W_p/\hbar$. Suppose that the system is next irradiated by frequency ω. In principle, although the set of frequencies $m\omega_p + n\omega$ could exist for all integer values of m and n, the system must be capable of supporting each of them for them to exist. If the system can support only the frequency ω_p, then only the two frequencies ω and ω_p will exist, and Eqs. (12-8) become

$$\frac{P_p}{\omega_p} = 0 = \frac{P}{\omega}, \tag{12-9}$$

so that no net power flows at either frequency.

Next, suppose that another energy state with energy W_q and corresponding frequency ω_q can exist, and that

$$\omega_q = a\omega_p + b\omega, \tag{12-10}$$

where a and b are integers. The combination of Eqs. (12-8) and (12-10) shows that

$$\begin{aligned} \frac{P_p}{\omega_p} + \frac{aP_{ab}}{a\omega_p + b\omega} &= 0, \\ \frac{P}{\omega} + \frac{bP_{ab}}{a\omega_p + b\omega} &= 0. \end{aligned} \tag{12-11}$$

The power flow, as shown in Fig. 12-1, goes from the original state p to the final state q such that

$$\frac{P}{\omega} = \frac{b}{a}\frac{P_p}{\omega_p}. \tag{12-12}$$

We see from Eq. (12-12) that power will flow from the irradiating field to the final state if a and b are of the same sign, i.e., absorption will occur, and power will flow from the original state to the irradiating field if a and b are of opposite sign, i.e., emission will occur.

Upon taking the power flowing into the mixing system as positive, the energy flowing out from the original state is that initially stored there, or

$$W_p = \int P_p(t)\, dt = \hbar\omega_p. \tag{12-13}$$

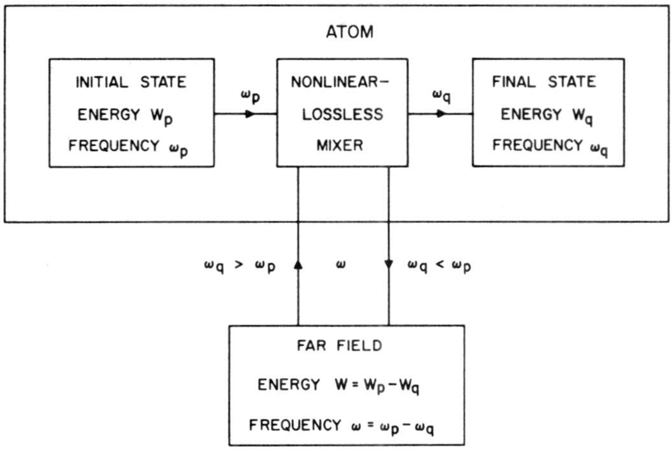

FIG. 12-1. Model for application of the Manley–Rowe equations to atomic radiation.

The energy exchanged with the far field follows from Eqs. (12-12) and (12-13):

$$W = \frac{b}{a}\hbar\omega.$$

The set of integers $a = 1$, $b = \pm 1$ is a typical example. For $b = 1$, $\omega_q = \omega_p + \omega$ and energy is absorbed from the irradiating field, whereas for $b = -1$, $\omega_q = \omega_p - \omega$ and energy is emitted to the irradiating field. The energy exchanged is

$$W = \hbar\omega. \tag{12-14}$$

The exchange of energy between the far field and a single atom with its possible eigenenergies, therefore, is quantized in units proportional to the frequency and to $\hbar$. Because this conclusion applies only to exchanges of energy between the background field and the atom, only field-atom interactions involving energy exchanges accompanying transitions between eigenstates are subject to the condition of Eq. (12-14). One exempt phenomenon is the shift of electron energy levels under the influence of perturbing radiation, which produces a nonunit relative permittivity; another is the nonlinearities that result in harmonic generation when an atom is subjected to extreme fields such as those present in laser beams.

The combination of Eqs. (7-21) with (12-14) shows that

$$L_z = \pm \hbar, \tag{12-15}$$

i.e., the angular momentum carried by the radiation, and therefore the change of angular momentum within the radiating region upon absorption or emission of electromagnetic energy, is $\hbar$.

The photoelectric effect is the name given the phenomenon from which, when electromagnetic radiation irradiates an object, electrons are ejected. The maximum energy of the ejected electron is in proportion to the frequency of the radiation. The number of ejected electrons is in proportion to the intensity of the radiation. There is no noticeable time delay between onsets of radiation and ejected electrons. These facts have been taken as evidence for the corpuscular theory of radiation. But the discussion accompanying Eq. (7-20) shows that an individual atom can absorb from incident radiation both the required energy and momentum. From the concept developed in Chapters 6–9 of each atom situated in the near-field of its neighbors, it follows that the total energy of an individual electron and the field need not be conserved, and Eq. (12-14) shows the absorbed energy to be proportional to frequency. Therefore, the phenomenon of photoelectricity is entirely consistent with observation without invoking a corpuscular nature to light.

REFERENCES

Einstein, Albert, "Concerning a heuristic point of view about the creation and transformation of light," *Ann. Physik.* **17,** 132–148 (1905). *Reprinted in* "World of the Atom" (H. A. Boorse, and L. Motz, eds.), pp. 544–557. Basic Books, New York, 1966.

Grimes, D. M., "Energy level transitions," *Proc. IEEE* **53,** 1663–1664 (1965).

Manley, J., and Rowe, H. E., "Some general properties of nonlinear elements. Part I: General energy relations," *Proc. IRE* **44,** 904–913 (1956).

Manley, J., and Rowe, H. E., "General energy relations," *Proc. IRE* **47,** 2115–2116 (1959).

Weiss, M. T., "Quantum derivation of energy relations analogous to those for nonlinear reactances," *Proc. IRE* **45,** 1012 (1957).

13
THE HARMONIC POTENTIAL

Solutions of the Schrödinger equation depend, of course, upon the detailed form of the potential as a function of position. One relatively simple and particularly relevant case is that of a harmonic oscillator. The solution is found directly from the Schrödinger equation:

$$-\frac{\hbar^2}{2m}\frac{d^2U(x)}{dx^2} + (\tfrac{1}{2}kx^2 - W)U(x) = 0. \tag{13-1}$$

To solve this equation, we first write it in dimensionless form, then separate out the asymptotic behavior for large values of x, and finally solve the resulting equation by means of a series expansion.

By writing $z = x/x_0$ and $U'' = d^2U/dz^2$, Eq. (13-1) can be rewritten as

$$U'' + \left(\frac{2Wmx_0{}^2}{\hbar^2} - \frac{mkz^2x_0{}^4}{\hbar^2}\right)U = 0. \tag{13-2}$$

Then, if we substitute $x_0{}^4 = \hbar^2/km$ and $\lambda = 2\ W/\hbar\omega_0$, where $\omega_0 = (k/m)^{1/2}$, the equation becomes

$$U'' + (\lambda - z^2)U = 0. \tag{13-3}$$

The chosen values of x_0 and λ in no way limit the solution, but the value of x_0 determines the scale factor.

The asymptotic form can be separated only by considering values of U when $z \gg \lambda$; the differential equation becomes approximately

$$U'' - z^2 U = 0, \tag{13-4}$$

which, for large values of z, is solved by putting

$$U(z) = \exp(\pm z^2/2). \tag{13-5}$$

Equation (13-5) describes the limiting behavior of U for very large values of the argument; so the total function can be expressed in the form

$$U(z) = N \exp(-z^2/2) H(z), \tag{13-6}$$

where $H(z)$ varies more slowly with z than does $\exp(-z^2/2)$, for large values of z, and where the sign of the exponent must of necessity be negative, for, if it were positive, the total charge would be infinite. Substituting Eq. (13-6) back into Eq. (13-4) and simplifying gives

$$H'' - 2zH' + (\lambda - 1)H = 0. \tag{13-7}$$

Equation (13-7) can be solved by a power series expansion. When the expanded form of $H(z)$ is written as

$$H(z) = \sum_{s=0}^{\infty} b_s z^s, \tag{13-8}$$

the differential equation becomes

$$\sum_{s=0}^{\infty} [s(s-1)b_s z^{s-2} + (\lambda - 1 - 2s)z^s b_s] = 0,$$

or, if the values of the dummy indices are shifted,

$$\sum_{s=0}^{\infty} [(s+2)(s+1)b_{s+2} + (\lambda - 1 - 2s)b_s]z^s = 0. \tag{13-9}$$

Since Eq. (13-9) must be satisfied for all values of z, each coefficient multiplying a power of z must be zero; that is,

$$b_{s+2} = \frac{1 + 2s - \lambda}{(s+1)(s+2)} b_s. \tag{13-10}$$

From Eq. (13-10), the solution for $H(z)$ can be completely known if one even coefficient and one odd one, say b_0 and b_1, are known.

We note from Eq. (13-10) that, in the limit of large values of s, the ratio of coefficients b_{s+2}/b_s is

$$\frac{b_{s+2}}{b_s} = \frac{2}{s}.$$

Since this is also the ratio of coefficients for $\exp(z^2/2)$, it would seem that the function $H(z)$ would increase as $\exp(z^2/2)$, and that the value of $U(z)$ would be constant. But this would result in an infinite total charge, and would violate the condition that $H(z)$ must change functionally much more slowly with distance than does $\exp(-z^2/2)$ for large values of the argument; hence the conclusion cannot be correct. It can be avoided, and Eq. (13-10) retained, only if

$$\lambda = 2n + 1, \tag{13-11}$$

where n is an integer. Therefore, the power series giving $H(z)$ terminates when $s = n$, since all coefficients with subscripts higher than n are 0. Similarly, the coefficient b_0 must also be zero for n odd, as must b_1 for n even. $H_n(z)$ is therefore a polynomial of order n and is either an even or an odd function of position, but is not a combination.

Equation (13-11) combines with the definition of λ to give

$$W_n = \hbar\omega_0\left(\frac{n+1}{2}\right). \tag{13-12}$$

The functions $H_n(z)$ are tabulated in many places. They are the Hermite polynomials, and they satisfy the integral equation

$$\int H_m(z)H_n(z)\exp(-z^2)\,dz = \delta_{mn}\,2^n n!\pi^{1/2} \tag{13-13}$$

and can be found from the differential relationship

$$H_n(z) = (-1)^n \exp(z^2)\,\frac{d^n \exp(-z^2)}{dz^n}. \tag{13-14}$$

The first few of the Hermite polynomials are listed in Table 13-1.

The product of a Hermite polynomial and the asymptotic form $\exp(-\ z^2/2)$ is a wave function; the first few are listed in Table 13-2.

TABLE 13-1
HERMITE POLYNOMIALS

$H_0(z) = 1$	$Hz(z) = 8z^3 - 12z$
$H_1(z) = 2z$	$Hz(z) = 16z^4 - 48z^2 - 12$
$H_2(z) = 4z^2 - 2$	$Hz(z) = 32z^5 - 160z^3 - 120z$

TABLE 13-2
HARMONIC POTENTIAL WAVE FUNCTIONS

$$\mathrm{U}_0(x) = [x_0\pi^{1/2}]^{-1/2} \exp\left(\frac{-x^2}{x_0^2}\right)$$

$$\mathrm{U}_1(x) = \left[\frac{2}{x_0\pi^{1/2}}\right]^{1/2} \frac{x}{x_0} \exp\left(\frac{-x^2}{x_0^2}\right)$$

$$\mathrm{U}_2(x) = [2x_0\pi^{1/2}]^{-1/2} \left(2\frac{x^2}{x_0^2} - 1\right) \exp\left(\frac{-x^2}{x_0^2}\right)$$

14

A SPHERICAL POTENTIAL

The time-independent Schrödinger equation for an electron in the vicinity of a spherically symmetric potential $I(r)$ produced by an infinite mass is

$$\frac{\hbar^2}{2m}\nabla^2 U(\mathbf{r}) + [-I(r) + W]U(\mathbf{r}) = 0, \tag{14-1}$$

and we seek properties of the wave function which depend upon the spherical nature of the potential. To do so, we begin by separating Eq. (14-1) into two separate equations, by writing $U(\mathbf{r})$ as a product function of radial and angular functions, just as we did with Eq. (5-8). The result of doing so is

$$R^{-1}\frac{d}{dr}\left(r^2\frac{dR}{dr}\right) + \frac{2mr^2}{\hbar^2}[-I + W]$$

$$= -(Y\sin\theta)^{-1}\frac{\partial}{\partial\theta}\left(\sin\theta\frac{\partial Y}{\partial\theta}\right) - (Y\sin^2\theta)^{-1}\frac{\partial^2 Y}{\partial\phi^2}.$$

And, similar to the previous analysis, all radial dependence appears on the left side of the equation, and all angular dependence on the right side. Therefore, each side is constant. As before, we put that constant equal to $l(l+1)$ to obtain the two separated equations

$$\frac{\hbar^2}{2mr^2}\frac{d}{dr}\left(r^2\frac{dR}{dr}\right)+\left[-I+W-\frac{l(l+1)\hbar^2}{2mr^2}\right]R(r)=0, \tag{14-2}$$

$$(\sin\theta)^{-1}\frac{\theta}{\partial\theta}\left(\sin\theta\frac{\partial Y}{\partial\theta}\right)+(\sin^2\theta)^{-1}\frac{\partial^2 Y}{\partial\phi^2}+l(l+1)Y=0. \tag{14-3}$$

Equation (14-2), the differential equation satisfied by the radial component of the wave function, is directly dependent upon the form of the potential. $R(r)$ will, therefore, depend upon the potential. The angular equation, however, does not explicitly depend upon the potential. We wish, then, to examine results of solving Eq. (14-3) to determine results that are independent of the detailed form of the spherical potential.

The arguments used in Chapter 5 to show that n (here called l) and $\boldsymbol{m}$ were integers are equally applicable here. Therefore, as before, n is a positive integer, and $\mathbf{m}$ is an integer that lies within the range

$$-l \leqslant \mathbf{m} \leqslant l.$$

The magnetic moment $\mathbf{M}$ and the angular momentum $\mathbf{L}$ produced by an electron in motion (as distinguished from the self values of Chapter 8) are given by

$$\begin{aligned}\mathbf{M}&=-\frac{1}{2}\frac{e}{m}\int U^*(\mathbf{r})U(\mathbf{r})\mathbf{r}\times\mathbf{g}(\mathbf{r})\,dV,\\ \mathbf{L}&=\int U^*(\mathbf{r})U(\mathbf{r})\mathbf{r}\times\mathbf{g}(\mathbf{r})\,dV.\end{aligned} \tag{14-4}$$

Now from Eq. (10-1) we see that the rms value of velocity is obtained by replacing

$$U^*U\mathbf{v} \qquad \text{by} \qquad i\frac{\hbar}{m}U^*\nabla U.$$

We therefore make the same substitution in Eqs. (14-4) to obtain

$$\mathbf{M} = -\frac{ie\hbar}{2m}\int U^*\mathbf{r} \times \nabla U\, dV,$$
$$\mathbf{L} = i\hbar \int U^*\mathbf{r} \times \nabla U\, dV. \qquad (14\text{-}5)$$

The operator $\mathbf{r} \times \nabla U$ can be evaluated by direct, although lengthy, algebra. The result is

$$(\mathbf{r} \times \nabla U)_x = -\sin\phi \frac{\partial U}{\partial \theta} - \operatorname{ctn}\theta \cos\phi \frac{\partial U}{\partial \phi},$$
$$(\mathbf{r} \times \nabla U)_y = \cos\phi \frac{\partial U}{\partial \theta} - \operatorname{ctn}\theta \sin\phi \frac{\partial U}{\partial \phi}, \qquad (14\text{-}6)$$
$$(\mathbf{r} \times \nabla U)_z = \frac{\partial U}{\partial \phi}.$$

When Eqs. (14-6) are substituted into Eqs. (14-5), and the orthogonality of the Legendre polynomials is utilized, the integrated result is $L_x = 0$, $L_y = 0$, and

$$L_z = \mathbf{m}\hbar,$$
$$M_z = -\frac{e\hbar\mathbf{m}}{2m}. \qquad (14\text{-}7)$$

The square of the angular momentum is

$$L^2 = \int U^*U(\mathbf{r} \times \mathbf{g})^2\, dV, \qquad (14\text{-}8)$$

where $(\mathbf{r} \times \mathbf{g})^2 = (\mathbf{r} \times \mathbf{g}) \cdot (\mathbf{r} \times \mathbf{g})$. Equations (14-6) combine with Eq. (14-8) to give

$$L^2 = \hbar^2 \int U^*\left[(\sin\theta)^{-1}\frac{\partial}{\partial\theta}\left(\sin\theta\frac{\partial Y}{\partial\theta}\right) + (\sin^2\theta)^{-1}\frac{\partial^2 Y}{\partial\phi^2}\right] dV. \qquad (14\text{-}9)$$

The combination of Eqs. (14-3) and (14-9) gives

$$L^2 = l(l+1)\hbar^2. \qquad (14\text{-}10)$$

Similarly, the square of the magnetic moment is

$$M^2 = l(l + 1)\left(\frac{e\hbar}{2m}\right)^2. \tag{14-11}$$

As we discussed in Chapter 8, an electron possesses an intrinsic spin and magnetic moment as well as an electric charge. The charge distributions as discussed by the Fourier transform technique of Chapter 9 include the dynamic charge distribution, but they do not include the intrinsic spin and magnetic moment of the electron. To include the spin, it is convenient to begin with the known values of spin σ and magnetic moment μ, collinear with an applied magnetic field:

$$\sigma_z = \mathbf{m}_s\hbar, \qquad \mu_z = -\, e/m\sigma_z, \tag{14-12}$$

where $\mathbf{m}_s = \pm 1/2$. The possible values of magnetic moment due to the spin are, then,

$$\mu_z = \pm \frac{e\hbar}{2m}. \tag{14-13}$$

The value of Eq. (14-13) is called the Bohr magneton. To place the spin-moment in a parallel scheme with the value of angular spin and moment, we introduce the square of the spin as

$$\sigma^2 = s(s + 1)\hbar^2, \tag{14-14}$$

where $s = 1/2$.

The z component of the spin and magnetic moment are the only values that it is possible to measure. These values have a fixed value along a line collinear with an applied magnetic field, as would be expected from Eq. (17-24). The measureable portion of the magnetic moment is that portion collinear with an applied field. It is convenient to consider Eqs. (14-10), (14-11), and (14-14) as the total value of the corresponding variable, and that the spins are distributed at a resulting angle with respect to the applied field which gives the correct value for the z component.

The parameter $\mathbf{m}_s$ is the fourth quantum number, and each wave function obtained from solutions to the Schrödinger equation is considered to be multiplied by an additional parameter whose eigenvalues show a spin of $\pm \hbar/2$.

15

THE COULOMB POTENTIAL

In Chapter 14, we considered results for a potential function on which the only restriction was that it be spherically symmetric. Let us turn now to a particular spherical potential: the Coulomb potential field for an electron in the vicinity of a nuclear charge of Ze coulombs. For this case, $\Phi(r) = -Ze/(4\pi\varepsilon r)$; putting this value into Eq. (14-1), the differential equation of motion for the radial component, gives the wave function. For this purpose, it is convenient to substitute

$$\zeta^2 = -\frac{8mW}{\hbar^2},$$

$$\lambda = \frac{Ze^2}{4\pi\varepsilon\hbar}\left(-\frac{m}{2W}\right)^{1/2} \tag{15-1}$$

and to substitute $z = \zeta r$ as the dynamic variable. For this case, Eq. (14-1) becomes

$$\frac{d^2R}{dz^2} + \frac{2}{z}\frac{dR}{dz} + \left[\frac{\lambda}{z} - \frac{1}{4} - \frac{l(l+1)}{z^2}\right]R = 0. \tag{15-2}$$

Equation (15-2) is in dimensionless form. It is most easily solved by retracing the steps taken in Chapter 13 to solve the equation for the harmonic potential. The first step is to remove that part of the differential equation which describes the asymptotic behavior for large values of the variable. In the limit of very large values of z, the differential equation becomes simply

$$\frac{d^2R}{dz^2} - \frac{R}{4} = 0,$$

an equation whose formal solution is, nearly,

$$R(z) = Fe^{-z/2} + Ge^{+z/2}.$$

If the total charge density is to remain integrable, i.e., finite over all space, the coefficient G must be zero. The differential equation of interest, however, is Eq. (15-2); so F is a function of z which for large distances varies much less rapidly with distance than does $e^{-z/2}$.

If we substitute $R(z) = F(z)e^{-z/2}$, and simplify somewhat, the complete differential equation becomes

$$\frac{d^2F}{dz^2} + \left(\frac{2}{z} - 1\right)\frac{dF}{dz} + \left[\frac{\lambda - 1}{z} - \frac{l(l+1)}{z^2}\right]F = 0. \tag{15-3}$$

Equation (15-3) is in a form such that it can be solved by using a power series expansion, i.e., by putting

$$F(z) = \sum_{s=0}^{\infty} c_s z^{t+s},$$

where t is chosen so that the first coefficient c_0 is different from zero. Substituting back into Eq. (15-3), and gathering terms, gives the summation

$$0 = z^{t-2}c_0[t(t+1) - l(l+1)] + \sum_{s=0}^{\infty} z^{s+t-1}\{c_s(b - 1 - t - s)$$

$$+ c_{s+1}[(t+s+1)(t+s+2) - l(l+1)]\},$$

which must be satisfied for all values of z. For this to be the case, each coefficient multiplying a power of z must equal zero. From the coefficient of the $t-2$ power of z, it follows that $t = l$, and the summation gives the recursion relationship,

$$\frac{c_{s+1}}{c_s} = \frac{1 + l + s - \lambda}{(s + l + 1)(s + l + 2) - l(l+1)}. \tag{15-4}$$

For large values of s, the ratio of the coefficients is $1/s$. But this is the ratio of coefficients for the infinite series expansion of the function e^z. In the limit of a very large argument, therefore, $F(z)$ will increase much more rapidly with distance than the asymptotic form, if the series continues. This clearly violates the condition imposed in the preceding paragraph, and will result in an infinite total charge; so the series must terminate. It will do so only if λ is an integer, let us say n, so that $F(z)$ will be a polynomial of highest order n, and also if

$$n > l. \tag{15-5}$$

Substituting Eq. (15-5) into the expression for b, Eq. (15-1), shows that the time-average energy of the electron is

$$W = -\frac{mZ^2e^4}{32\pi^2\varepsilon^2n^2\hbar^2}. \tag{15-6}$$

Since we shall be using Eq. (15-6) in several different ways, let us rewrite it in several forms. For $h = 2\pi\hbar$, it is

$$W = -\frac{mZ^2e^4}{8\varepsilon^2n^2h^2}, \tag{15-7}$$

and if we introduce the dimensionless fine-structure constant α, where

$$\alpha = \frac{1}{2}\left(\frac{\mu}{\varepsilon}\right)^{1/2}\frac{e^2}{h} \cong (137.0373)^{-1}, \tag{15-8}$$

it can be written as

$$W = -\alpha^2(mc^2)\frac{Z^2}{2n^2}.$$

Note that the ergodic value of energy is quantized, as it was for the harmonic oscillator.

The functions $F(z)$ are related to the Laguerre polynomials, and the first few are listed in Table 15-1. The polynomials are functions of both quantum numbers, n and l, and are listed as F_{nl}. To obtain the total wave function, the functions $F(z)$ must be multiplied by the asymptotic form $e^{-z/2}$ extracted earlier,

TABLE 15-1

THE POLYNOMIALS $F(z)$

$F_{10}(z) = 1$	$F_{30}(z) = 1 - z + z^2/6$	$F_{40}(z) = 1 - 3z/2 + z^2/2 - z^3/24$
$F_{20}(z) = 1 - z/2$	$F_{31}(z) = z - z^2/4$	$F_{41}(z) = z - z^2/2 + z^3/20$
$F_{21}(z) = z$	$F_{32}(z) = z^2$	$F_{42}(z) = z^2 - z^3/6$
		$F_{43}(z) = z^3$

and by the proper spherical harmonic. The average value of distance between nucleus and electron can be found by direct integration over the wave functions. The average value of $1/r$ can be found directly from the energy, by using the virial theorem. For the Coulomb potential, the potential energy is twice the kinetic energy; therefore,

$$I(r) = -\frac{Ze^2}{4\pi\varepsilon}\langle r^{-1}\rangle = -\frac{mZ^2e^4}{4\varepsilon^2n^2h^2}$$

or

$$\langle r^{-1}\rangle = \alpha\frac{Zmc}{\hbar n^2} = b^{-1}\frac{Z}{n^2}, \tag{15-9}$$

where b is the Bohr orbit of the electron,

$$b = \frac{\hbar}{\alpha mc}. \tag{15-10}$$

The complete set of wave functions for the first three values of n is listed in Table 15-2. For wave functions with $l = 0$, a plot of the charge versus distance would show a maximum at $r = b$. For $l > 0$, the maximum occurs at a larger value of r. For the $m = 0$ wave function, the effective charge density is stationary; for $m \neq 0$, the charge density itself is rotating about the nucleus so as to produce the angular momentum and magnetic moment given in Eqs. (14-7).

TABLE 15-2
COULOMB POTENTIAL WAVE FUNCTIONS U_{nlm}

$$U_{100} = (\pi b^3)^{-1/2} e^{-r/b}$$

$$U_{200} = (32\pi b^3)^{-1/2} \left(2 - \frac{r}{b}\right) e^{-r/2b}$$

$$U_{210} = (32\pi b^3)^{-1/2} \frac{r}{b} e^{-r/2b} \cos\theta$$

$$U_{21\pm1} = (64\pi b^3)^{-1/2} \frac{r}{b} e^{-r/2b} \sin\theta\, e^{\pm i\phi}$$

$$U_{300} = (3\pi b^3)^{-1/2} \frac{1}{81}\left(27 - \frac{18r}{b} + \frac{2r^2}{b^2}\right) e^{-r/3b}$$

$$U_{310} = \left(\frac{2}{\pi b^3}\right)^{1/2} \frac{1}{81}\left(6 - \frac{r}{b}\right)\frac{r}{b} e^{-r/3b} \cos\theta$$

$$U_{31\pm1} = \left(\frac{2}{\pi b^3}\right)^{1/2} \frac{1}{81}\left(6 - \frac{r}{b}\right)\frac{r}{b} e^{-r/3b} \sin\theta\, e^{\pm i\phi}$$

$$U_{320} = (6\pi b^3)^{-1/2} \left(\frac{r}{b}\right)^2 e^{-r/3b} (3\cos^2\theta - 1)$$

$$U_{32\pm1} = (\pi b^3)^{-1/2} \left(\frac{r}{b}\right)^2 e^{-r/3b} \sin\theta\cos\theta\, e^{\pm i\phi}$$

$$U_{32\pm2} = (\pi b^3)^{-1/2} \left(\frac{r}{b}\right)^2 e^{-r/3b} \sin^2\theta\, e^{\pm 2i\phi}$$

16

TRANSITION PROBABILITIES

Let us now examine possible transitions within the radiating region, and the probability of an electron's undergoing a transition between eigenstates of an atom. To do so, we need detailed information on the structure of the atom. The Manley–Rowe equations, like thermodynamic equations, are general enough to enable one to derive important conclusions without stipulating a detailed mechanism: in Chapter 12, we used them to show that perturbing electromagnetic radiation can be absorbed or emitted only in the energy units stated by Eq. (12-14). But this same generality precludes their use in obtaining detailed information on the atom itself. For our present purpose, therefore, we must use the detailed structural characteristics of the atom.

By Eq. (11-7), the time-dependent wave function is

$$\psi(t, \mathbf{r}) = \sum_a C_a U_a(\mathbf{r}) \exp\left(\frac{iW_a t}{\hbar}\right), \qquad (16\text{-}1)$$

where the running index a indicates a sum over all possible eigenstates and where $C_s{}^*C_s$ is the probability that the electron charge density is given by $U_s{}^*(\mathbf{r})U_s(\mathbf{r})$. We begin by assuming that the atom is within an externally applied electrostatic field. The Hamiltonian for the atom in isolation is H_0, and the presence of the field increases it by H_1. Let us restrict the problem to those cases in which $\hat{H}_0\psi \gg \hat{H}_1\psi$, and assume that the field increases from zero to its final value. The interaction term is always

δH_1, where δ is a dimensionless number that varies from zero to unity as a function of the strength of the interaction.

The wave function U and the energy W undergo shifts with δ which can be written in power series expansions, thus:

$$\begin{aligned} W &= W_0 + \delta W_1 + \delta^2 W_2 + \cdots, \\ U &= U_0 + \delta U_1 + \delta^2 U_2 + \cdots, \end{aligned} \tag{16-2}$$

where $W_i \gg W_{i+1}$. Putting Eqs. (16-2) into the Schrödinger equation gives

$$\begin{aligned} 0 = (\hat{H}U_0 - W_0 U_0) + \delta(\hat{H}U_0 - W_0 U_1 + \hat{H}U_1 - W_1 U_0) \\ + \delta^2 \cdots + \cdots, \end{aligned}$$

and, since δ can be controlled independently, the coefficient of each power of δ must be zero. By Schrödinger's equation, the term independent of δ sums to zero. The first-order correction to the interacting systems is described by the factor proportional to δ. Higher-order terms will be ignored.

Multiplying the coefficient of δ by U^* and integrating over all space gives

$$0 = \int U_0^*(\hat{H}_1 - W_1)U_0 \, dV + \int U_0^*(\hat{H}_0 U_1 - W_0 U_1) \, dV. \tag{16-3}$$

From Eq. (10-3), the second integral is

$$\int U_1^*(\hat{H}_0 - W_0)U_0 \, dV,$$

where the integrand is proportional to the Schrödinger equation, and so is equal to zero. This leaves the new result

$$\int U_0^*(\hat{H}_1 - W_1)U_0 \, dV = 0. \tag{16-4}$$

Equation (16-4) enables one to find the results with knowledge of only the unperturbed wave functions, so long as the interaction remains small. That is, the wave functions, along with the corresponding resonant frequencies, remain the same as those obtained in the absence of an interaction energy, and the individual wave functions $U_a(\mathbf{r})$ of Eq. (16-1) remain unchanged during the interactions. The coefficients C_a, however, although formerly constant, are now time-dependent. In other words, the interaction energy causes the distribution probabilities among the possible eigenstates to be time-dependent. To find this time dependence, let us substitute Eq. (16-1) directly into the time-dependent Schrödinger equation

$$(\hat{H}_0 + \hat{H}_1)\psi(t, \mathbf{r}) = \frac{\hbar}{i}\frac{\partial\psi}{\partial t} = W\psi, \tag{16-5}$$

with the stipulation that $C_a = C_a(t)$, to obtain

$$\sum_a C_a(t)(\hat{H}_0 + \hat{H}_1)\psi_a = \frac{\hbar}{i}\frac{\partial}{\partial t}\sum_a C_a(t)\psi_a$$

or

$$\sum_a \left(C_a\hat{H}_0\psi_a + C_a\hat{H}_1\psi_a - \frac{\hbar}{i}\psi_a\frac{dC_a}{dt} - C_a\frac{\hbar}{i}\frac{\partial\psi_a}{\partial t}\right) = 0. \tag{16-6}$$

Since the first and last terms of this equation sum to zero, we obtain the new result

$$\sum_a \left(C_a\hat{H}_1\psi_a - \frac{\hbar}{i}\psi_a\frac{dC_a}{dt}\right) = 0. \tag{16-7}$$

Equation (16-7) can be further simplified by using the orthogonality of the wave functions ψ_a. Multiplying through by $\psi_b{}^*$ and integrating over all space gives

$$\frac{dC_b}{dt} = \frac{i}{h}\sum_a (b \mid \hat{H}_1 \mid a)C_a \exp\frac{it}{\hbar}(W_a - W_b), \tag{16-8}$$

where

$$(b \mid \hat{H}_1 \mid a) = \int U_b{}^*\hat{H}_1 U_a\, dV. \tag{16-9}$$

For H_1 independent of time, and for $W_a \neq W_b$, the time-average value of Eq. (16-8) will be zero. If the electric field is time-dependent, then the interaction energy is given by

$$H_1 = -e\mathbf{v}\cdot\mathbf{A}, \tag{16-10}$$

where **A** is the magnetic vector potential in the perturbing radiation field and **v** is the velocity of the charge. We assign the frequency ω to the radiation field, and, from Eq. (2-12), $\mathbf{E} = -i\omega\mathbf{A}$; so Eq. (16-8) becomes

$$\frac{dC_b(t, \omega)}{dt} = \frac{e}{\omega\hbar}\sum_a (b \mid \mathbf{v} \cdot \mathbf{E}_0 \mid a)C_a \exp\left[\frac{it}{\hbar}(W_a - W_b \pm \hbar\omega)\right], \tag{16-11}$$

where $\mathbf{E} = \mathbf{E}_0 e^{\pm i\omega t}$, and where C_b is a function of both time and frequency. We now integrate Eq. (16-11) to obtain $C_b(t, \omega)$. We take the initial conditions that $C_s = 1$ and $C_a = 0$ for $a \neq s$; then

$$\frac{dC_b(t, \omega)}{dt} = \frac{e}{\hbar\omega}(b \mid \mathbf{v} \cdot \mathbf{E}_0 \mid s) \exp\left[\frac{it}{\hbar}(W_s - W_b \pm \hbar\omega)\right],$$

and

$$C_b(t, \omega) = -\frac{e}{i\omega}(b \mid \mathbf{v} \cdot \mathbf{E}_0 \mid s)\left[\frac{\exp\left[\frac{it}{\hbar}(W_s - W_b \pm \hbar\omega)\right] - 1}{W_s - W_b \pm \hbar\omega}\right]$$

with the result that the probability of state b being occupied is

$$C_b{}^*(t, \omega)C_b(t, \omega) = \frac{4e^2}{\omega^2}(b \mid \mathbf{v} \cdot \mathbf{E}_0 \mid s)^2 \frac{\sin^2 \frac{t}{2\hbar}(W_s - W_b \pm \hbar\omega)}{(W_s - W_b \pm \hbar\omega)^2}.$$

Now, if external radiation of frequency ω brings about a change of eigenstate, then

$$C_b{}^*(t)C_b(t) = \int_{-\infty}^{\infty} \frac{4e^2}{\omega^2}(b \mid \mathbf{v} \cdot \mathbf{E}_0 \mid s)^2 \times \left[\frac{\sin^2 \frac{t}{2\hbar}(W_s - W_b \pm \hbar\omega)}{(W_s - W_b \pm \hbar\omega)^2}\right]\rho(\omega)\, d\omega, \tag{16-13}$$

where $\rho(\omega)$ is the intensity of the radiation at the atom and it has a frequency between ω and $\omega + d\omega$, i.e., it is the density of possible radiation frequencies or states. Now we put $z = (t/h)(\hbar\omega \pm W_s \mp W_0)$ and note that $z^{-1} \sin z$ is appreciably different from zero only when $z = 0$. We define $\omega_0 = \pm \hbar^{-1}(W_b - W_s)$ and note that Eq. (16-13) can be rewritten as

$$C_b^* C_b = \frac{2e^2 \rho(\omega_0)}{\omega_0^{\,2}} (b \,|\, \mathbf{v} \cdot \mathbf{E}_0 \,|\, s)^2 \frac{t}{\hbar^2} \int_{-\infty}^{\infty} \frac{\sin^2 (z/2)}{z^2} \, dz$$

$$= \frac{2\pi e^2 \rho(\omega_0)}{\hbar^2 \omega_0^{\,2}} (b \,|\, \mathbf{v} \cdot \mathbf{E}_0 \,|\, s)^2 t. \qquad (16\text{-}14)$$

As can easily be shown from electromagnetic theory, the density of radiation states $\rho(\omega)$ in a cubical conducting box of sides L is

$$\rho(\omega) = \frac{L^3 \omega^2}{\pi^2 c^3}. \qquad (16\text{-}15)$$

And one requirement, in agreement with Eq. (12-14), is

$$W = W_b - W_s = \pm \hbar\omega. \qquad (16\text{-}16)$$

According to Eq. (16-14), the probability of transition increases in proportion to t. However, this is so only because we assumed $C_s = 1$ at all times. A more exact solution would produce a time dependence similar to that for a well-known classical example, in which two lossless parallel resonant circuits are close enough together so that the inductive coupling is small, but not negligible. If one circuit is energized but the other is not, and if both resonate at the same frequency, all energy from the first circuit will be transferred to the second, and then back again. In a lossless system, the process will continue indefinitely. The magnitude of the coupling coefficient will determine the frequency with which the energy is transferred, but not the amount transferred. Nor, in the particularly interesting case of low coupling, will it materially affect the resonant frequencies of the circuits. In quantized systems, the time dependence of the probability of electron exchange between two eigenstates is very similar. If the electron is first in state b, it will be transferred to state s, then back to state b, etc.

The term $(b|\mathbf{v} \cdot \mathbf{E}_0|s)$ simplifies considerably if the wavelength of the radiation field is very much larger than the size of the atom. For that case, $\mathbf{E}_0$ can be removed from the integrand to obtain

$$(b \mid \mathbf{v} \cdot \mathbf{E}_0 \mid s) = \mathbf{E}_0 \cdot \int U_b^* \mathbf{v} U_s \, dV.$$

Combining this with Eq. (11-10) gives

$$(b \mid \mathbf{v} \cdot \mathbf{E}_0 \mid s) = \frac{i}{\hbar} \mathbf{E}_0 \cdot \int U_b^* [\mathbf{r} \hat{H} U_s - \hat{H}(\mathbf{r} U_s)] \, dV.$$

Combining this, in turn, with Eq. (10-3) gives

$$(b \mid \mathbf{v} \cdot \mathbf{E}_0 \mid s) = \frac{i}{\hbar} \mathbf{E}_0 \cdot \int [U_b^* \mathbf{r}(\hat{H} U_s) - \mathbf{r} U_s^*(\hat{H} U_b)] \, dV.$$

And, finally, combining this with Eq. (10-5) gives

$$(b \mid \mathbf{v} \cdot \mathbf{E}_0 \mid s) = \frac{i}{\hbar} (W_s - W_b) \mathbf{E}_0 \cdot (b \mid \mathbf{r} \mid s). \tag{16-17}$$

The product $-e(b|\mathbf{r}|s)$ is the dipole moment produced by the electron as it oscillates between states b and s.

Let us evaluate $(b|\mathbf{r}|s)$ for a single electron in an attractive, spherically-symmetric electrostatic potential field. The integral is the sum of three independent ones, i.e., one for each of the orthogonal directions x, y, and z. Shifting to primed and unprimed notation to distinguish between wave functions enables us to write these three integrals as

$$X_i = \int U^*_{n'l'm'} x_i U_{nlm} \, dV,$$

where

$$U_{nlm} = R_n(r) P_l^m(\cos\theta) e^{-im\phi},$$

in which spin is ignored. The integrals for the three orthogonal directions are, respectively,

$$X = \tfrac{1}{2} \int R_{n'} R_n r^3 \, dr \int \sin^2\theta \, P_{l'}^{m'} P_l^m \, d\theta \int e^{-i(m-m')\phi}(e^{i\phi} + e^{-i\phi}) \, d\phi, \tag{16-18}$$

$$Y = \tfrac{1}{2} \int R_{n'} R_n r^3 \, dr \int \sin^2\theta \, P_{l'}^{m'} P_l^m \, d\theta \int e^{-i(m-m')\phi}(e^{i\phi} - e^{-i\phi}) \, d\phi, \tag{16-19}$$

$$Z = \int R_{n'} R_n r^3 \, dr \int \sin\theta \cos\theta \, P_{l'}^{m'} P_l^m \, d\theta \int e^{-i(m-m')\phi} \, d\phi, \tag{16-20}$$

Each of the three terms X_i is, in turn, the product of three integrals. From these integrals, it follows that the matrix element for the transition is zero unless $\Delta l = \pm 1$ and unless $m = \pm 1$ or 0 by using Eq. (16-21). These are the only dipolar selection rules, since the integral $Q(n', n)$ imposes no restriction. The following identities with respect to associated Legendre polynomials are useful:

$$(2l + 1) \sin\theta \, P_l^{m-1}(\cos\theta) = P_{l+1}^{m}(\cos\theta) - P_{l-1}^{m}(\cos\theta),$$

$$(2l + 1) \sin\theta \, P_l^{m+1}(\cos\theta) = (l + m + 1)(l + m)P_{l-1}^{m}(\cos\theta) - (l - m)(l - m + 1)P_{l+1}^{m}(\cos)\,\theta),$$

$$(2l + 1) \cos\theta \, P_l^{m}(\cos\theta) = (l - m + 1)P_{l+1}^{m}(\cos\theta) + (l + m)P_{l-1}^{m}(\cos\theta). \qquad (16\text{-}21)$$

We have ignored electron spin in developing these selection rules. Since selection rules have been treated in detail in many texts, we shall not go into their development here; suffice it to say that the selection rules, including spin effects, are

$$\begin{aligned} \Delta l &= \pm 1, \\ \Delta s &= 0, \\ \Delta j' &= \pm 1 \text{ or } 0, \\ \Delta m_j &= \pm 1 \text{ or } 0, \end{aligned} \qquad (16\text{-}22)$$

with the additional restrictions that $\Delta j'$ cannot be zero if j' is zero and $m_j' = 0$ cannot go to another state for which $n_{j'} = 0$ if $\Delta j'$ is zero.

In summary, the perturbing field produces oscillations between eigenstates separated by energy $\hbar\omega$ in the direction, and at the frequency, of the perturbing radiation. These oscillations either emit to the field or absorb energy reversibly from the field until the energy condition of Eq. (12-14) is met. From that point on, the electron oscillates about the new equilibrium position until the requirement $E = \hbar\omega$ is again met, etc. We note, in particular, that the probability of an electron's absorbing radiation and going to a higher energy state is the same as the probability of its emitting radiation and going to a lower energy state.

17

ANGULAR TORQUE ON DYNAMIC SYSTEMS

It was shown in Chapter 7 that an atomic system excited by a plane wave will, under the proper circumstances, radiate or absorb power and, in the general case, accompanying momentum and angular momentum as well. Equations (7-7), (7-21), and (12-15) are expressions that formally describe the exchanged angular momentum. Now, if angular momentum is to be radiated away from the source, then the angular momentum of the source must be changed by the negative of that amount. And that source momentum can change only if the source itself is subject to a field-generated torque. That torque is described in the first part of this chapter.

An atomic model was developed in Chapter 8. That model consists of an electron, about a positive nucleus, that is in dynamic equilibrium between the attractive electrostatic force of the nucleus and the driving force of the reactive field generated by the electron's own motion. Based upon this model, only the far-field, power-carrying terms sum to zero. Near-field and inductive field effects persist. In the latter portion of this chapter, the torque that results from the interaction of the remaining generated terms with an applied, uniform, static magnetic field is calculated. The result is seen to be a time-dependent torque that rotates the atomic moment to be either parallel or anti-parallel with the externally applied field. This effect is known as space quantization; i.e., the average moments are collinear with an applied field and are not disturbed over a continuous range of angles.

The calculation of the torques begins with the model for the source of a perfectly conducting very small sphere carrying surface currents that generate the fields of Eqs. (5-33). We wish to find the torque on the sphere of radius $a \ll \lambda$ produced by the action of all external fields on the surface currents. To do so, we apply Eq. (4-27) over the source region. Since the Poynting's vector remains finite on the spherical surface, the remaining terms are

$$E_j + (\mathbf{J} \times \mathbf{B})_j = \frac{\partial \sigma_{jk}}{\partial x_k}, \tag{17-1}$$

and the torque, by definition, is $\mathscr{T} = \mathbf{r} \times \mathbf{F}$. Thus the applicable torque equation on the source is

$$\mathscr{T}_i = \int dV(x_j F_k^{\text{v}} - x_k F_j^{\text{v}}) = \oint (x_j \sigma_{kn} - x_k \sigma_{jn})\, dS_n, \tag{17-2}$$

where dS_n is the nth component of the source's surface area and, in spherical coordinates, is everywhere directed in the radial direction and equal in magnitude to

$$dS_n = a^2 \sin\theta\, d\theta\, d\phi = a^2 d\Omega. \tag{17-3}$$

The differential torque on each differential surface area is, then,

$$d\mathscr{T} = a^3(\sigma_{\theta n}\hat{\phi} - \sigma_{\phi r}\hat{\theta})\, d\Omega,$$

which, in rectangular coordinates, is

$$\begin{aligned} d\mathscr{T} = a^3\, d\Omega[&-\hat{x}(\sigma_{\theta r}\sin\phi + \sigma_{\phi r}\cos\theta\cos\phi) \\ &+ \hat{y}(\sigma_{\theta r}\cos\phi - \sigma_{\phi r}\cos\theta\sin\phi) \\ &+ \hat{z}\sigma_{\phi r}\sin\theta]. \end{aligned} \tag{17-4}$$

To find the total torque, $\sigma_{\theta r}$ and $\sigma_{\phi r}$ must be evaluated and then the result integrated over the spherical surface of the source region. The σ_{ij} are evaluated by using Eqs. (2-18). Since we seek the average values, we take one-half of the real part of the product between one field and the complex conjugate of the other. The result of combining Eqs. (2-18) and (5-33) in that way is

$$\sigma_{\theta r} = \mu n(n+1)\frac{h_n(x)}{2x}\,[(a_{nm}a^*_{nm} + b_{nm}b^*_{nm})H_n{}^* Y_n{}^m Q_n^{m*}$$

$$- i(a_{nm}b^*_{nm} + a^*_{nm}b_{nm})h_n{}^* Y_n{}^m S_n^{m*}], \tag{17-5}$$

$$\sigma_{\phi r} = \mu n(n+1)\frac{h_n{}^*(x)}{2x}[- i(a_{nm}a^*_{nm} + b_{nm}b^*_{nm})H_n Y_n^{m*} S_n{}^m$$

$$+ (a_{nm}b^*_{nm} + a^*_{nm}b_{nm})h_n Y_n^{m*} Q_n{}^m],$$

where $h_n(x)$ is any normalized, linear combination of Hankel functions of the first and second kind.

We consider first the case of either emission or absorption. Then $h_n(x)$ represents a spherical Hankel function of the second or first kind, respectively. From Chapter 7, we know that induced radiation can occur only on those terms for which $m = \pm 1$, and, for those cases, the combination of Eqs. (17-4) and (17-5) integrated over the azimuthal angle ϕ shows that

$$\mathscr{T}_x = 0 = \mathscr{T}_y, \tag{17-6}$$

$$\mathscr{T}_z = a^3\mu n(n+1)\frac{h_n{}^*(x)}{2x}\oint d\Omega \sin\theta[- i(a_{nm}a^*_{um} + b_{nm}b^*_{nm})$$

$$\times H_n Y_n^{m*} S_n{}^m + (a_{nm}b^*_{nm} + a^*_{nm}b_{nm})h_n Y_n^{m*} Q_n{}^m]. \tag{17-7}$$

The surface integrals of Eq. (17-7) have been evaluated in Eq. (7A-9) and, for the case when only one value of n exists, in Eq. (7A-10). The torque is, then,

$$\mathscr{T}_z = - ia^3\mu n(n+1)m\frac{h_n{}^*(x)}{2x}H_n(x)(a_{nm}a^*_{nm} + b_{nm}b^*_{nm}). \tag{17-8}$$

If no power is exchanged between the source and the far field, then both h_n and H_n are real, and so no real torque exists. If power is exchanged, on the other hand, then the torque is proportional to the real part of the product $ih_nH_n{}^*$. For $n = 1$, after using Table 5-2 and the definition of $H_n(x)$, the resultant torque on the source is

$$\mathscr{T}_z = \pm \frac{\eta}{ck^3} \sum_m m(a_{1m}a^*_{1m} + b_{1m}b^*_{1m}). \tag{17-9}$$

Comparison of Eqs. (7-7) and (17-9) shows that the rate at which angular momentum is transferred to the far-field differs from the torque on the source produced by the field only by the plus or minus sign—a result that has to be obtained if angular momentum is to be conserved.

Next let us consider the torque on a nonradiating source that is immersed within a uniform externally applied static magnetic field, $\mathbf{B}_0$. The relationship between the torque and the fields is, once again, given by Eqs. (17-4). We seek the instantaneous value of the torque. There is no loss of generality upon adjusting the coordinate system so B_0 lies within the X–Z plane:

$$\mathbf{B}_0 = B_0(\hat{z} \cos \theta_0 + \hat{x} \sin \theta_0). \tag{17-10}$$

The total magnetic field is the sum of the field terms from Eqs. (5-33) and (17-10). These terms are, after making the substitution that $\mu H_0 = B_0$,

$$\begin{aligned} H_r &= n(n+1)b_{nm}\frac{h_n(x)}{x} Y_n{}^m e^{i\omega t} + H_0(\cos\theta\cos\theta_0 \\ &\quad + \sin\theta\cos\phi\sin\theta_0), \\ H_\theta &= ia_{nm}h_n S_n{}^m e^{i\omega t} + b_{nm}H_n Q_n{}^m e^{i\omega t} + H_0(-\sin\theta\cos\theta_0 \\ &\quad + \cos\theta\cos\phi\sin\theta_0) \qquad\qquad (17\text{-}11) \\ H_\phi &= a_{nm}h_n Q_n{}^m e^{i\omega t} - ib_{nm}H_n S_n{}^m e^{i\omega t} - H_0\sin\phi\sin\theta_0. \end{aligned}$$

The electric field components of the stress tensor σ_{ij} do not interact with the field, and so they will yield no net torque. Those terms in σ_{ij} which involve $H_0{}^2$ give the torque due to the applied field acting upon itself, whereas those product components not involving H_0 give the torque due to the region acting upon itself. All such terms integrate to zero. We are, therefore, concerned only with those product terms in σ_{ij} which are proportional to H_0:

$$\begin{aligned}
2\sigma_{\theta r} &= B_0(ia_{nm}h_nS_n{}^m + b_{nm}H_nQ_n{}^m)(\cos\theta\cos\theta_0 \\
&\quad + \sin\theta\cos\phi\sin\theta_0)e^{i\omega t} \\
&\quad + B_0 n(n+1)b_{nm}\frac{h_n}{x}Y_n{}^m(-\sin\theta\cos\theta_0 \\
&\quad + \cos\theta\cos\phi\sin\theta_0)e^{i\omega t}, \\
2\sigma_{\phi r} &= B_0(a_{nm}h_nQ_n{}^m - ib_{nm}H_nS_n{}^m)(\cos\theta\cos\theta_0 \\
&\quad + \sin\theta\cos\phi\sin\theta_0)e^{i\omega t} \\
&\quad - B_0 n(n+1)b_{nm}\frac{h_n}{x}Y_n{}^m\sin\phi\sin\theta_0 e^{i\omega t}.
\end{aligned} \tag{17-12}$$

The torque on each differential area is found by combining Eqs. (17-4) and (17-12):

$$\begin{aligned}
2\frac{d\mathcal{T}_x}{dS} &= -aB_0e^{i\omega t}\{(\cos\theta\cos\theta_0 + \sin\theta\cos\phi\sin\theta_0) \\
&\quad \times[(ia_{nm}h_nS_n{}^m + b_{nm}H_nQ_n{}^m)\sin\phi \\
&\quad +(a_{nm}h_nQ_n{}^m - ib_{nm}H_nS_n{}^m)\cos\theta\cos\phi] \\
&\quad - n(n+1)b_{nm}\frac{h_n}{x}Y_n{}^m\sin\theta\sin\phi\cos\theta_0\}, \\
2\frac{d\mathcal{T}_y}{dS} &= aB_0e^{i\omega t}\{(\cos\theta\cos\theta_0 + \sin\theta\cos\phi\cos\theta_0) \qquad (17\text{-}13) \\
&\quad \times[(ia_{nm}h_nS_n{}^m + b_{nm}H_nQ_n{}^m)\cos\phi \\
&\quad -(a_{nm}h_nQ_n{}^m - ib_{nm}H_nS_n{}^m)\cos\theta\sin\phi] \\
&\quad + n(n+1)b_{nm}\frac{h_n}{x}Y_n{}^m(\cos\theta\sin\theta_0 - \sin\theta\cos\phi\cos\theta_0)\}, \\
2\frac{d\mathcal{T}_z}{dS} &= aB_0e^{i\omega t}[(\cos\theta\cos\theta_0 - \sin\theta\cos\phi\sin\theta_0)\sin\theta\,(a_{nm}h_nQ_n{}^m \\
&\quad - ib_{nm}H_nS_n{}^m) - n(n+1)b_{nm}\frac{h_n}{x}Y_n{}^m\sin\theta\sin\phi\sin\theta_0].
\end{aligned}$$

Although Eq. (17-13) is rather long, it simplifies considerably for each of several special cases. We consider first the case of $m = 0$, and put $a_n = a_{n0}$ and $b_n = b_{n0}$. For these cases, the terms that do not integrate to zero over the azimuthal angle are

$$2\frac{d\mathcal{T}_x}{dS} = -aB_0 a_n h_n Q_n \sin\theta\cos\theta\cos^2\phi\sin\theta_0 e^{i\omega t},$$

$$2\frac{d\mathcal{T}_y}{dS} = aB_0 e^{i\omega t}(b_n H_n Q_n \sin\theta\cos^2\phi\sin\theta_0 + n(n+1)b_n \frac{h_n}{x} Y_n \cos\theta\sin\theta_0), \tag{17.14}$$

$$2\frac{d\mathcal{T}_z}{dS} = aB_0 a_n h_n Q_n \sin\theta\cos\theta\cos\theta_0\, e^{i\omega t}.$$

Now the function Y_n is an even function of θ, and Q_n is an odd function of θ if n is even, and the converse is true if n is odd. Therefore, referring back to Eq. (17-14), $\mathcal{T}_x$ and $\mathcal{T}_z$ integrate to zero for n odd but not for n even, whereas $\mathcal{T}_y$ integrates to zero for n even but not for n odd. Furthermore, $\mathcal{T}_x$ and $\mathcal{T}_z$ satisfy the relationship

$$\mathcal{T}_z = -2\,\mathrm{ctn}\,\theta_0 \mathcal{T}_x. \tag{17-15}$$

The angular-directed torques, in the general case, are related to the foregoing torques calculated in the special case of a field applied in the X–Z plane by

$$\mathcal{T}_\theta = -\mathcal{T}_x \sec\theta_0, \qquad \mathcal{T}_\phi = \mathcal{T}_y. \tag{17-16}$$

For the $n = 1$ case, the angular functions are

$$Y_1 = \left[\frac{3}{4\pi}\right]^{1/2} \cos\theta_0, \qquad Q_1 = -\left[\frac{3}{4\pi}\right]^{1/2} \sin\theta_0, \tag{17-17}$$

and the result of combining Eqs. (17-14), (17-16), and (17-17) and integrating over the angles is

$$\mathscr{T}_\phi = \left[\frac{4\pi}{3}\right]^{1/2} b_1 a^3 B_0 \sin\theta_0 \left[\frac{h_1(ka)}{(ka)^2} - H_1(ka)\right] e^{i\omega t}. \tag{17-18}$$

Equation (17-18) can be simplified for a small source by taking the limit as ka becomes much less than one and by expressing b_1 in terms of the Z-directed magnetic moment m

$$b_1 = -\frac{mk^3}{(12\pi)^{1/2}}, \tag{17-19}$$

with the result that

$$\mathscr{T}_\phi = mB_0 \sin\theta_0 e^{i\omega t} \tag{17-20}$$

Equation (17-20) is, of course, the usual torque expression for a magnetic dipole m in an external magnetic field B_0. The torque remains nonzero for zero frequency excitation.

The next case is for $n = 2$, where

$$\begin{aligned} Y_2 &= \frac{1}{2}\left(\frac{5}{4\pi}\right)^{1/2}(3\cos^2\theta - 1), \\ Q_2 &= -3\left(\frac{5}{4\pi}\right)^{1/2} \sin\theta\cos\theta. \end{aligned} \tag{17-21}$$

The nonzero angular torque $\mathscr{T}_\theta$ is found, by using Eqs. (17-14), (17-16), and (17-21), to be

$$\mathscr{T}_\theta = -\left(\frac{4\pi}{5}\right)^{1/2} B_0 a_2 h_2 \tan\theta_0 e^{i\omega t}, \tag{17-22}$$

or the torque is proportional to the electric quadropolar moment. If p_2 is the quadropolar moment along the Z axis, then

$$a_2 = \frac{icp_2k^4}{6(20\pi)^{1/2}}, \tag{17-23}$$

and, for a small source, the real part of the resultant torque is

$$\mathscr{T}_\theta = -\frac{p_2 B_0}{10} \tan\theta_0 \sin\omega t. \tag{17-24}$$

Next we let I be the moment of inertia of the source. Then, by definition of torque and the moment, the source satisfies the equation

$$I\frac{d^2\theta_0}{dt^2} = \frac{p_2 B_0}{10}\tan\theta_0 \sin\omega t. \tag{17-25}$$

Equation (17-25) cannot be readily solved in terms of elementary functions. However, if $\theta_0 < \pi/2$ at the onset of the magnetic field, then, because of the singularity in $\tan\theta_0$ at $\theta_0 = \pi/2$, the magnetic moment will remain within the range $0 \leqslant \theta_0 < \pi/2$. On the other hand, if $\theta_0 > \pi/2$ at the onset of the field, then the magnetic moment will remain within the range $\pi/2 < \theta_0 \leqslant \pi$. Now if θ_0 does not equal either 0 or π, then the direction of the moment will oscillate under the influence of the applied field, and as it oscillates it will radiate energy. Such a nonzero damping coefficient will cause the moment to equilibrate collinear with the direction of the field.

Gerlach and Stern first showed experimentally that electronic moments are collinear with an applied field, as expressed by Eq. (14-13). This effect is known as space quantization, and is often said to be a purely quantum-mechanical one.

The result of Eq. (17-25) is that the axis of the time-varying electronic quadropole moment is rotated to be collinear with the magnetic field. Now if the source also supports a magnetic dipolar moment collinear with its electric quadropolar axis, then the magnetic moment is collinear with the field, and the result is space quantization.

REFERENCE

Gerlach, W., and Stern, O., "Experimental proof of space quantization in a magnetic field," *in* "World of the Atom" (H. A. Boorse, and L. Motz, eds.), pp. 936–939. Basic Books, New York, 1956.

SUBJECT INDEX

A

a_{nm}, b_{nm} coefficients, 67
Aberration, *see* Light
Absorption, 2, 53, 68, 87, 114, 116, 137
Acceleration, 12, 30, 33, 34, 51
Ampere's law, 31
Angular momentum, 51, 68, 75, 76, 78–80, 116, 122, 123, 128, 139, 142
  radiated, 63
Angular velocity, 84
Antennas, 59–62
  miniaturized, 70
Aperture A_e, 61
Atomic physics, 1–5
Azimuthal angle, 141, 144

B

Babylonians, theory of, 4
Bandwidth, 59
Bessel functions, spherical, 65
Bloch, theory of, 3
Bohr, theory of, 2, 3
Bohr magneton, 124
Bohr orbit, 128
Born, theory of, 3

C

Capacitor, 111–113
Cartesian coordinates, 43
Cauchy integral, 25
Central limit theorem, 91
Charge, 75
  configurations, 80
  density of, *see* Charge density
  distribution, 84, 85
  static, 36
  velocity of, 133
Charge density, 16, 36, 77, 81, 83, 84, 88, 91, 94, 99, 101–103, 107, 126, 128, 131
Charge q, 15, 31, 54, 107, 111
Complex conjugate, 89, 90, 98, 108, 140
Compton wavelength, 76, 78
Configuration force, 56
Continuity equation, 107, 108
Coordinate-space function, 97
Coordinates, *see* specific kinds
Cosmological principle, 8, 15
Coulomb potential, 125–129
  wave functions, 129
Coulomb's law, 31
Coupling coefficients, 135
Current density, 81–84, 107
  root-mean-square, 88
Current density $\mathbf{J}$ $(t, \mathbf{r})$, 80
Cylindrical coordinates, 43

D

Damping coefficient, 146
Decay time τ, 59
Differential volume, 101, 102

Diffusion constant, 94
Dipole moment, 37, 56, 99, 136, 146
Dirac delta function, 23, 26
Distribution probabilities, 133

E

Eigenenergies, 115
Eigenstates, 115, 125, 133–136
 of atom, 131
Eigenvalues, 124
Einstein, theory of, 2, 3, 4, 69
Electric charge, 15, 16, 18, 25, 29, 124
 conservation of, 108
Electric dipole, 39, 53
Electric field, 30, 31, 37, 39, 41, 42, 49, 133, 142
Electric field vectors, 18, 66
Electric multipoles, 51–53
Electric quadrupole, 56
Electricity, 1
Electrodynamics, 18, 29–40
Electromagnetic field, 76
Electromagnetic field theory, 75
Electromagnetic mass, 75
Electromagnetic radiation, 116
Electromagnetic tensors, 15–22
Electromagnetic theory, 135
Electrons, 75–85, 87
 comparative lengths, 76
 diameter, 78, 79
 mass, 78
 time-average energy of, 127
Electrostatic field, 131
 effect, 55
Electrostatic force, 139
Electrostatic potential, 36, 88
Electrostatic potential field, 105, 136
Electrostatic potential functions, 100, 101
Emission, 2, 53, 69, 87, 114, 116, 137
Energy, 35
 absorption, 69
 flow of, 51, 63–73
 of interaction, *see* Interaction energy
 negative binding, 76
 radiant, 111
 radiated, 63
 reactive, 57, 58, 62, 67
 shifts, 132
 storage, 51–62
Energy-to-momentum ratio, 67
Equilibrium, 79, 80, 87, 88, 137, 139
Ergodic equation, 87–95
 time-dependent, 107–110

F

Field point, 29, 36, 81
Field vectors, 29, 82
Fitzgerald contraction, 10
Flow, in multipolar fields, 51–62
Force
 on charge, 110
 laws, 29
Four-current density, 16
Four-force, 17
Four-potential, evaluation of, 23–27
Four-space, 24, 26
Four-surface, 32
Four-tensors, 16–20
Four-vectors, 11–13, 16–20, 24, 25
Four-velocity, 13, 17, 18, 32
Four-volume, 18, 32
Fourier transform, 24, 80, 83, 88, 124
 inverse, 82
Frequency, 108, 113, 114, 116, 134
 minimum, 57
 resonant, 132

G

Gauss's law, 94
Green's function, 23, 29

H

Hamiltonian, 98, 131
Hankel functions, 48, 52, 141
  spherical, 49
Hanson's method, 42
Harmonic oscillator, 57, 117, 127
Harmonic potential, 117–120, 125
  wave functions, 120
Heisenberg, theory of, 3
Heisenberg uncertainty principle, 103, 105
Hermite polynomials, 119, 120
Hermitian operators, 98, 109
Hermitian properties, 105

I

Impedance, 52, 53, 61
Inertia, moments of, *see* Moments of inertia
Inertial effect, 55
Inertial systems, 15
Interaction energy, 102, 132, 133

K

Kinetic energy, 89, 128

L

Laguerre polynomials, 127, 128
Laplacian operator, 43
Laplacian vector, 41, 42
Lasers, 116
Legendre polynomials, 65, 70, 123
  associated, 46, 48, 137
Lienard-Wiechert potentials, 30
Light, aberration of, 13
Lorentz covariant tensors, 15
Lorentz diameter, 76
Lorentz model, 35
Lorentz transformations, 7–15, 20
Low-velocity limit, 30

M

Magnetic dipole, 39, 145
Magnetic field, 30, 31, 38, 39, 41, 42, 49, 78, 124, 139, 142, 145, 146
Magnetic field vectors, 18, 66
Magnetic moment, 39, 75, 77, 95, 122, 124, 128, 145, 146
Magnetic multipoles, 52
Magnetic vector potential, 133
Magnetism, 1
Manley-Rowe equations, 113, 115, 131
Mass, 34, 35, 75, 107
  of particles, 76
Mass density, 76
Maxwell equations, 3, 18, 41, 59
Maxwell stress tensor, 20, 39
Moment of inertia, 146
Moments
  dipole, *see* Dipole moments
  magnetic, *see* Magnetic moments
  quadrupole, *see* Quadrupole moment
Momentum, 51, 69, 70, 105, 139
  angular, *see* Angular momentum
  flow, 63–73
  radiated, 63
Momentum density, 33, 34, 107
Motion, 8–10
  equation of, 56
Multipolar fields, 41–73
Multipole radiators, 54

N

Neutrons, 75
Newton's laws, 7, 31, 33, 110

O

Orthogonal set, 99
Oscillator, harmonic, *see* Harmonic oscillator

P

Parity, 100
Particles, *see also* Electrons, Protons
  size, 78
Pauli exclusion principle, 103
Perturbation theory, 132
Photoelectric effect, 116
Photon, radiation pattern for, 69
Planck radiation law, 2
Point charge, 1, 29, 56
Potential, harmonic, *see* Harmonic potential
Potential energy, 56, 58, 89, 128
Potential field, 99
Power, radiated, 54
  effect, 55
Power absorption, 52, 62, 68, 69, *see also* Absorption
Power density, 35
Power emission, 52, 62, 68, 69, *see also* Emission
Power flow, 51, 57, 111, 112, 114
Power loss, by radiation, 58, 61
Poynting vector, 19, 36, 61, 63, 64, 140
Protons, 75
  comparative lengths, 76
  diameter, 78
  mass, 78
Pulsations, radial, 85

Q

Quadrupole moment, 145, 146
Quantum-mechanical operators, 105

R

Radial components of electric and magnetic fields, 43, 66
Radial pulsations, *see* Pulsations
Radian frequency, 68, 94
Radiation, 8–11
  corpuscular theory of, 116
  density of, 135
  dipolar, 56–58
  electric multipolar, 67
  impedance, 60
  induced, 141
  magnetic dipolar, 67
  patterns, 60
  quantum theory of, 69
Radiation field, 133, 134
Radiation frequencies, 134
Radiation reaction, 56
Radiation resistance, 61
Radiating atom, 111–116
Radiative Q, 58
Random walk, theory of, 94
Reciprocity theorem, 59, 60
Resonant circuits, 135

S

Schrödinger equation, 3, 91, 97, 98, 100, 113, 117, 124, 132, 133
  time-dependent, 108
  time-independent, 94, 121
Schwarz inequality, 104
Source point, 29
Source velocity, 8
Space
  concept of, 7
  three-dimensional, *see* Three-space
Space quantization, 139, 146
Spherical coordinates, 43
Spherical harmonics, 70
Spherical potential, 121–124

Spherical shell, 83–84
Spin, 84, 85, 124
  effects of, 137
Stationary system, 10, 11
Surfaces, four-dimensional, 18

T

Taylor series, 37, 92
Three-space, 18, 27, 30
Three-tensor, 20
Time
  concept of, 7
  interval, 10
  proper, 11
Torques, 39
  angular, on dynamic systems, 139–146
  calculation of, 140
  field-generated, 139
Transition probabilities, 131–137

V

Vector potential, 38, 81
Velocity, 12, 13, 30
  angular, *see* Angular velocity
  average, 93
  of charged particle, 91
Virial theorem, 128
Voltage, 111, 112

W

Wave function, properties of, 97–106

ELECTRICAL SCIENCE

A Series of Monographs and Texts

Editors

Henry G. Booker
UNIVERSITY OF CALIFORNIA AT SAN DIEGO
LA JOLLA, CALIFORNIA

Nicholas DeClaris
UNIVERSITY OF MARYLAND
COLLEGE PARK, MARYLAND

Joseph E. Rowe. Nonlinear Electron-Wave Interaction Phenomena. 1965

Max J. O. Strutt. Semiconductor Devices: Volume I. Semiconductors and Semiconductor Diodes. 1966

Austin Blaquiere. Nonlinear System Analysis. 1966

Victor Rumsey. Frequency Independent Antennas. 1966

Charles K. Birdsall and William B. Bridges. Electron Dynamics of Diode Regions. 1966

A. D. Kuz'min and A. E. Salomonovich. Radioastronomical Methods of Antenna Measurements. 1966

Charles Cook and Marvin Bernfeld. Radar Signals: An Introduction to Theory and Application. 1967

J. W. Crispin, Jr., and K. M. Siegel (eds.). Methods of Radar Cross Section Analysis. 1968

Giuseppe Biorci (ed.). Network and Switching Theory. 1968

Ernest C. Okress (ed.). Microwave Power Engineering:
Volume 1. Generation, Transmission, Rectification. 1968
Volume 2. Applications. 1968

T. R. Bashkow (ed.). Engineering Applications of Digital Computers. 1968

Julius T. Tou (ed.). Applied Automata Theory. 1968

Robert Lyon-Caen. Diodes, Transistors, and Integrated Circuits for Switching Systems. 1969

M. Ronald Wohlers. Lumped and Distributed Passive Networks. 1969

Michel Cuenod and Allen E. Durling. A Discrete-Time Approach for System Analysis. 1969

K. Kurokawa. An Introduction to the Theory of Microwave Circuits. 1969

H. K. Messerle. Energy Conversion Statics. 1969

George Tyras. Radiation and Propagation of Electromagnetic Waves. 1969

Georges Metzger and Jean-Paul Vabre. Transmission Lines with Pulse Excitation. 1969

C. L. Sheng. Threshhold Logic. 1969

Dale M. Grimes. Electromagnetism and Quantum Theory. 1969

Robert O. Harger. Synthetic Aperture Radar Systems: Theory and Design. 1969

In Preparation

M. A. Lampert and P. Mark. Current Injection in Solids.

W. V. T. Rusch and P. D. Potter. Analysis of Reflector Antennas.